High energy costs

Uneven, unfair, unavoidable ?

High energy costs

Uneven, unfair, unavoidable ?

Hans H. Landsberg and Joseph M. Dukert

PUBLISHED FOR RESOURCES FOR THE FUTURE, INC.
BY THE JOHNS HOPKINS UNIVERSITY PRESS, BALTIMORE AND LONDON

Published for Resources for the Future
By The Johns Hopkins University Press, Baltimore, Maryland 21218

ISBN 0-8018-2781-7 (hardcover)
     0-8018-2782-5 (paperback)

Library of Congress Cataloging in Publication Data

Landsberg, Hans H.
   High energy costs—uneven, unfair, unavoidable?

   Includes index.
   1. Energy consumption—United States—Costs—Congress.
I. Dukert, Joseph M.   II. Resources for the Future.   III. Title
HD9502.U52L347   333.79′17   81-15648
ISBN 0-8018-2782-5   AACR2

Resources for the Future is a nonprofit organization for research and education in the development, conservation, and use of natural resources and the improvement of the quality of the environment. It was established in 1952 with the cooperation of the Ford Foundation. Grants for research are accepted from government and private sources only if they meet the conditions of a policy established by the Board of Directors of Resources for the Future. The policy states that RFF shall be solely responsible for the conduct of the research and free to make the research results available to the public. Part of the work of Resources for the Future is carried out by its resident staff; part is supported by grants to universities and other nonprofit organizations. Unless otherwise stated, interpretations and conclusions in RFF publications are those of the authors; the organization takes responsibility for the selection of significant subjects for study, the competence of the researchers, and their freedom of inquiry.

This book is a product of RFF's Center for Energy Policy Research, Milton Russell, director. Hans H. Landsberg is a senior fellow in the division, and Joseph M. Dukert is a consultant specializing in energy. The conference on which this book is based was sponsored jointly by RFF and the Brookings Institution, with funding provided by the U.S. Department of Energy, the Office of Technology Assessment of the U.S. Congress, the Ford Foundation, and the Rockefeller Foundation.

The book was edited by Jo Hinkel and designed by Elsa B. Williams. The index was prepared by Barbara Bacon, and the figures were drawn by Federal Graphics.

Contents

Foreword

Increasingly, the United States has embraced the market as an instrument of energy policy with impressive results. More energy is reaching the market, and consumption has been reduced because of conservation measures. Those who use energy in consumption as well as in production processes have responded to higher prices by reducing their energy use to an extent greater than that predicted by many of the most ardent advocates of the market as a policy instrument.

These are all new dimensions in the U.S. energy experience. Prior to the early 1970s, energy costs were declining relative to other prices in the economy, whereas energy use was increasing overall, as well as on a per capita basis. Our institutions, technology, and expectations were geared toward producing abundant energy with declining costs.

It is not surprising, then, that we were nervous about letting the market for energy reflect a more curtailed supply. For, despite the assurances of various specialists, we really could not be certain that higher prices would bring more energy to the market, nor could we agree on the extent to which higher prices would stimulate energy conservation. But there was another reason for our cautious approach to pricing. No one knew how different groups would be affected by letting energy become relatively more expensive. Would the Northeast be at a disadvantage relative to the Southwest? Would the poor and the aged suffer, as compared with, say, those supplying energy? Would those who had energy to develop enjoy windfall gains unrelated either to their efforts or foresight?

Further, while it was clear that many would be affected in both positive and negative ways, it was unclear which effect would be the more important.

Resources for the Future has had a long history of research on energy policy and, in response to the intensity of national need, this research became more extensive during the 1970s. In the summer of 1979, two books appeared which offered powerful documentation for the positive contribution of the marketplace to the solution of energy problems. One, *Energy in America's Future: The Choices Before Us*, was published by The Johns Hopkins University Press for Resources for the Future. The other, *Energy: The Next Twenty Years*, published by the Ballinger Publishing Company, reported on a Ford Foundation study, conducted by a committee of experts from several disciplines that was chaired by Hans H. Landsberg of RFF, one of the authors of this volume.

The precise effect of the above books on public policy is unknown, although they have had excellent reviews in the scholarly and popular press and have been widely quoted both in political and industry circles.

But even with the publication of these books the question of impact remains, and it plagues our conscience. Have we achieved energy conservation at the expense of the less fortunate among us? Have some people benefited out of proportion to their contribution? Are we shifting the regional advantage within our nation in a fundamental, and possibly irreversible, way by the policies we are following? Several of us at RFF decided that we had an obligation to address some of these unanswered questions.

Consequently, Hans Landsberg and Milton Russell designed a conference to enable those having expertise in various aspects of the U.S. society to bring that knowledge to bear on such questions. Scholarly papers were presented, discussed, and criticized over a two-day period. The Proceedings of this conference, to be published later this year by RFF, no doubt will become a valuable resource for those in the scholarly community who are concerned about energy policy and income distribution.

Because of their technical nature, however, the papers in the Proceedings volume will not be as helpful to the busy congressman, the cabinet officer, or the concerned citizen. With this in mind, Hans Landsberg and Joseph M. Dukert undertook to make the conference findings accessible to anyone having an interest in the subject. I believe all readers of this present volume will agree that they have achieved this without sacrificing substance, an accomplishment that testifies to their skills in economics and communication.

RFF and the Brookings Institution organized the two-day conference, which was held at Brookings in late 1980. The funding for the conference and the subsequent manuscript preparation was generously provided by the U.S. Department of Energy, the Office of Technology Assessment of the U.S. Congress, the Ford Foundation, and the Rockefeller Foundation. Joseph Pechman of Brookings, and Allan G. Pulsipher, then of the Ford Foundation, deserve special recognition for representing their respective institutions so ably.

Washington, D.C.
September 1981

Emery N. Castle, President
Resources for the Future

Preface and acknowledgments

This publication is a hybrid—in more ways than one. It is the product of collaboration between an economist and a writer. It is a report on a conference, but supplemented by new material and reflecting the authors' judgments on choice of topics, on balance, and on scope. It is written with the clear purpose of making the material available to a broad readership, but it also strives to be useful to the scholar and the technician. It would be too much to expect that we have fully succeeded in each of these efforts, but we hope we have done so to a respectable degree.

Our thanks go in the first place to Emery Castle, who suggested that the topic needed investigation and that the time was now. As the project developed, we have had the benefit of critical review and useful suggestions from several colleagues at RFF, especially from Milton Russell, Joel Darmstadter, and Marion Clawson, and from Nathan Rosenberg of Stanford University. Jo Hinkel saw the book through editing and publication. Lee Carlson was responsible for the typing, as well as for many of the small but critical chores that go into planning a conference. Catherine T. McDonough and Sandra L. Glatt had the difficult task of verifying those statements and numbers in the text that lend themselves to the checking process.

To all of them go our thanks.

Washington, D.C.
September 1981

Hans H. Landsberg
Joseph M. Dukert

1
Stating the issues

As an energy-importing country with an energy-intensive economy, the United States is poorer than it was eight years ago—when the Arab oil embargo and a jarring rise in OPEC prices introduced us to the term *energy crisis*. The direct and indirect burdens that accompanied the end of the era of cheap, abundant, and reliable energy have fallen unevenly.

Inflation, joblessness, and intermittent shortages affect some groups, sectors, and regions more than others. A relative few have actually benefitted, while others have been squeezed from several directions at once. And many citizens who have finally accepted the unpleasant fact that our national energy problems cannot be resolved quickly or painlessly still ask—now almost plaintively: "Can't we insist at least that energy policy be *fair*? Shouldn't the costs imposed by energy's changing role be divided equally among all of us?"

Some limited but important steps toward answering such questions were taken late in 1980 at a two-day conference of approximately fifty persons in Washington, D.C., organized by Resources for the Future and the Brookings Institution. Its topic was "High Energy Cost: Assessing the Burden."[1] Participants ranged from congressional and federal executive branch staffers to public interest activists and academics. In ideology and affiliation they spanned the spectrum from economic and political conservatives to Marxists, from consumer advocates to public utility officers.

It would be naive to pretend that the discussions produced anything approaching unanimity about the concept, let alone the specific mechanics of combining efficient energy policy with the principles of social and economic equity. Nevertheless, Thomas C. Schelling, of the John F. Kennedy School of Government at Harvard University, was probably not exaggerating when he commented near the end of the first day that the conference had already "at least multiplied by two the amount of information available on this most important of subjects." Significantly, there were some areas of general agreement. Above all, there was basic consensus that energy prices should reflect full costs so that they would presumably continue to rise during the 1980s. Even participants at opposite ends of the spectrum concurred with the overwhelming majority of discussants who suggested in one way or another that the country can adjust to energy realities best by bringing energy prices close to true marginal replacement costs—despite the fact that moves in this direction can threaten those who live hand-to-mouth (whether they be individuals, families, business enterprises, or whole areas).

There are numerous ways of providing social assistance to the poor other than holding down energy prices

Such a judgment is not to be equated with lack of concern about the poor. Far from it. It is just that the conferees agreed almost universally that social equity problems and energy problems can and should be distinguished from one another. The two sets of problems are interrelated of course. The most recent available data presented at the conference showed, for instance, that the poorest Americans rank highest in the percentage of income spent on direct purchases of energy and are least capable of adopting many conservation measures. Yet there are numerous ways of providing social assistance to the poor other than holding down energy prices. A clear majority of the conferees seemed to believe that those "other ways" were more consistent with the national policy of promoting efficient use of energy.

Logic is not always a politically salable commodity, however. Many of those who took part in the conference were openly apprehensive that the conservative swing in public sentiment and in federal government policies, visible even in the preelection days of late 1980, would substantially reduce welfare efforts (and even regional or sectoral assistance programs) unless they were bound to the often dramatically demonstrable inequities arising from plainly visible and highly publicized rises in energy prices. It is still too soon to judge whether such fears are well-founded. Yet it is probably true, as one participant suggested, that "the general public has consistently perceived higher energy prices as the problem and not the solution." Except for the Marxist-oriented participants, who saw the discussion of "efficiency" and even "equity" as distractions and irrelevancies designed to perpetuate the power of the "managerial class" which they see as controlling U.S. energy policy, most conferees saw realistic energy pricing as beneficial on both sides of the supply-demand equation; it can encourage domestic production of energy and simultaneously discourage its inefficient use.

The general public has consistently perceived higher energy prices as the problem and not the solution

There is no reason why energy problems should be needed as either a crutch or an excuse for good social welfare policies. The same is true of what one might call "constructive federalism." If some segments of the U.S. population have been especially disadvantaged—regardless of the degree to which rising energy costs are involved—the decisions whether or not they should be helped by revenue sharing, income supplements, tax concessions, or aid in kind (such as food supplements) can certainly be divorced from a consideration of measures to reduce the nation's troublesome dependence on imported oil. Nevertheless, everything in the real world seems to be connected to everything else; and, even though the so-called Windfall Profits Tax may be an excise levy with no direct tie to profits, it certainly offers a windfall in federal revenues—which might be allocated to equalizing opportunities among groups in this country who are now unacceptably out of balance with each other. At least that is the judgment the Ninety-sixth Congress expressed when it passed the Windfall Profits legislation.

Although a brief conference must limit its scope, this one tried to take as little as possible for granted. For example, instead of accepting at face value the intuitive notion that price increases always hit the poor hardest, several conferees tested that hypothesis against fresh evidence from the Bureau of the Census and the Department of Energy's Energy Information Administration. While the new data tended to confirm the idea, discussion also uncovered a great many circumstances that qualify the rough overall impression. The poor are by no means homogeneous in energy needs, energy habits, or even the extent to which they rely on conventional income. In short, the problem of promoting equity (or even defining it) is far more complex than one might think.

The problem of promoting equity (or even defining it) is far more complex than one might think

On the other hand, the scope of the conference was deliberately limited in a number of important respects. It did not deal with the differential impact of environmental effects imposed by the production, transportation, distribution, and use of energy. It did not ask who bears the costs (for example, those in coal-mining areas or threatened coastal zones) and who enjoys the benefits (for example, those burning the fuel in boilers or automobile engines), or what adjustments can or should be made between them. Nor did it delve deeply into the impact of rising energy prices on the economy as a whole by way of inflation, depressed levels of aggregate economic activity, balance of payments problems, and so forth. Slowdowns in economic growth, productivity, or the special problems of energy-associated industries (for example, automobiles), all have their impact on income distribution, employment, and well-being, but they were not the focus of the undertaking. Finally, no attention was paid to the difference in impact on the domestic versus the international scene, and especially the "world poor" as compared with the domestic poor, and changes in income distribution on a worldwide scale. The subject matter was defined to set clear boundaries and produce a manageable set of issues. But the reader should be alerted to these limitations.

The purpose of this report is to review some of the facts, approaches, and proposals that surfaced during the conference—along with what turned out to be some blind alleys (such as looking for parallels in other energy-importing, industrialized countries, whose situations proved to be sufficiently different from that of the United States to be of only limited usefulness as models; a paper on foreign experiences will be published as part of the Proceedings). We think the facts, analyses, and comments, only occasionally supplemented by the authors of this report, deserve to be considered during the new round of national debate on energy policies which accompanies the change in administration and the legislative complexion of the country; but it is clear that such public discussions could be well advanced before a formal Proceedings volume is published.

This is not a consensus document. Essentially, it is an effort to distill from the commissioned papers, and the conference, a helpful picture of both the problem of rising energy costs and equity and the remedies—placing both in a broad context for continuing analysis and reflection. Like the conference itself, it is only a beginning.

Note

1. Conferees and the commissioned papers are listed in appendix B. Proceedings of the conference will be published at a later date.

2

Who pays the rising cost of energy?

Parallels may exist, but it is hard to point to any sustained increase in the price of a basic material as sudden and as steep as that of world crude oil during the period 1973–80. Just prior to the imposition of the oil embargo in late 1973, the posted price of Saudi Arabian oil was $3.01 per barrel at a Persian Gulf port of embarkation. Then it quadrupled, virtually in a single leap; and in 1979—after some intervening rises—it more than doubled again. By the end of 1980 it had climbed to $32, equivalent to about $19 in 1973 dollars.

Domestic energy prices have risen less steeply, but eventually all domestic oil will sell at world market prices. Further, the prices of other energy forms (natural gas, coal, and others) will adjust to the oil price on the basis of their comparative merits. President Reagan's administration was less than two weeks old when he moved to deregulate domestic oil eight months ahead of schedule, making an early down payment on the Republican party's commitment to free market principles. Interestingly, one of the final analytical studies of the outgoing Democratic administration's Department of Energy[1] contained remarkably similar preferences for adherence to market signals in all segments of energy. It counseled a speedup of the price decontrol timetable on natural gas, and introduction of a "security premium" on imported oil which could raise its price to U.S. consumers somewhere from $4 to $10 per barrel *above* the world level.

What does this shift mean to the individuals and families whose incomes are near or below the poverty line? To phrase it more generally and abstractly: Have energy price movements changed income and expenditure patterns in an undesirable direction? How does the shift affect a southwestern United States which has fashioned its own living and production style around an abundance of cheap natural gas, or a weatherbound New England whose local energy resources are virtually nonexistent? Could suddenly magnified commuting costs reverse the flight from central cities to suburbs that persisted in this country for more than a generation? Again, phrasing it more generally, we must ask: Where do the burdens of higher energy costs fall? What are their dimensions and nature? What tradeoffs are possible or desirable? Do the inequities that spring from adjustments violate our society's sense of fairness? Basically, in fact, what does "equity" mean?

Direct and indirect energy costs

Direct energy costs are not too difficult to compute. Between 1973 and 1980, raw energy of all sorts (oil, gas, coal, hydropower, and nuclear-generated electricity) more than tripled as a percentage of the U.S. gross national product. By the latter year, energy in its primary forms represented about 7 percent of the GNP[2]—and accounted for something like twice as much at the end-use level. Yet this far understates the full cost of energy for many . . . and the total effect of energy price changes on the country.

Increases in energy prices have not been uniform

Increases in energy prices have not been uniform—not by region, nor by source—as the next two chapters will spell out. Furthermore, while per-capita or per-household consumption of energy is a useful and necessary statistical device, even the amount and type of energy which might be termed an *essential minimum* vary so much as to make averages only vague indicators in gauging individual cost burdens.

Both the magnitude and the impact of indirect energy costs are far more difficult to assess. To begin with, higher costs of a production factor like energy necessitate a new balancing of factors—that is, a different mix of labor, capital, and energy. In the short run this is most likely to raise the costs of the end product and at times to adversely affect productivity. In addition, price–wage–price spirals tend to yield erratic results. The techniques of estimating the reflection of rising energy costs in other goods and services could stand refinement and updating; but the few studies undertaken suggest that the "embodied energy" used by an average household (for example, the energy that goes into producing, processing, and delivering food and other items) is slightly less than, but of the same order of magnitude as, the energy purchased directly in the form of heating fuel, power, gasoline, and so on.

One such estimate, using 1976 data, puts total indirect energy outlays by the very poor at 63 percent of direct payments for energy, while those with an income of around $20,000 were spending 85 percent as much on embodied energy as on direct energy. To the extent that this relationship can be accepted, it says something about the price effect. That is to say, the upward push on the Consumer Price Index from direct payments for energy might be doubled once the indirect energy purchases are included.

Exacerbating these straightforward price effects are macroeconomic problems such as unemployment and overall reductions in the rate of increase in GNP. Rising prices curb demand; lower consumption causes layoffs and business failures; where capacity is underused, lost advantages of scale may reduce productivity. At the conference, Gar Alperovitz, of the National Center for Economic Alternatives, quoted a House Commerce Subcommittee estimate of a loss of 800,000 jobs as a result of oil price decontrol, even if it had followed its original gradual timetable. And he quantified the cost to the nation by citing an OMB rough estimate, according to which each percentage point in increased unemployment is worth about $70 billion annually in lost national income and roughly $25 billion in uncollected federal tax revenues.[3]

Employment effects are easy to envision but tough to quantify. On the one hand, cutbacks in business travel resulting from higher fuel costs reduce the opportunities for commercial and industrial growth, yet the incentive for more efficient use of the telephone and teleprinter might eventually improve productivity. At some point, however, less travel (either for business or recreation) is likely to mean reduced economic opportunities in the hotel field—affecting maids and bellmen as well as managers and entrepreneurs. Such examples can be multiplied. On the other hand, replacing energy with labor would seem to create new employment opportunities. Indeed, Amory Lovins and others insist that shifting from an energy-intensive to a labor-intensive lifestyle would eventually produce jobs for more people; and so they would, by definition. Yet the social need is not just for jobs, but for jobs at a high enough

Employment effects are easy to envision but tough to quantify

wage rate to sustain current living standards. The employment-creation argument typically limits itself to the former.

If the consequences of a policy to accept continuing increases in the real price of energy seem unattractive, there may be some consolation in weighing them against the alternative. Continued underpricing of one or more forms of domestic energy is likely to discourage both efficient energy use and increased U.S. energy production. As a result the outlook for shifting toward greater self-sufficiency in energy mix would be dim, with the nation's fate indefinitely at the mercy of OPEC, or subject to the vagaries of supply from areas of the world where instability has been the rule rather than the exception in recent history, or both. Equally important, if we continue to underprice, then we misallocate resources. The whole economic pie shrinks and we pay through losses in real GNP. In sum, nobody regards rising energy costs as a favorable development (except perhaps in a very long-run ecological perspective, in which reduced energy consumption is seen as preserving the global environment). The costs to society as a whole are inevitable and varied. The more real resources we must invest in obtaining energy, either at home or from abroad, the less we will have left over to satisfy other wants. Moreover, there are a great many temporary perturbations to the economy, especially if price changes are rapid and large. It is as simple as that.

The debate during the seventies has been more over relative magnitudes than the direction of change: How large a bite will specified energy price increases take out of the GNP and living standards? How big a boost will energy price increases give to inflation? How many and whose jobs will be lost? And so on and so forth.

The conference did not focus on such calculations, though some participants offered their own or others' attempts at quantification. Emphasis was on where the burden falls, its total magnitude, how it is or ought to be shared, and the relationship of hardships associated with rising energy prices to poverty generally.

Equity: Meaning and measurement

This brings us back to the fact that "equity" itself remains undefined. As a general rule, Americans with higher incomes spend more on energy than those with lower incomes—absolutely, but not proportionately so. If the burden of higher costs is gauged in absolute terms, one concludes that the rich are hurt most. If we measure burdens in relation to the percentage of income—or spending power—that these costs represent, the poor emerge as the chief sufferers—and even more so if the marginal dollar is the basis of comparison. That marginal dollar has a much greater significance for the person with a low income than for one with a high income.

Furthermore, how does equity relate to waste? Oil, gas, gasoline, and electricity may be used extravagantly by anyone, regardless of where he lives, what he does, or what his income is. Does equity demand that some compensation be made for the fact that imprudent energy habits penalize a person more these days, when energy costs more?

Another related problem is that we lack a base period during which burdens were shared equitably. Instead, we have only base periods that were full of built-in inequities, of the kind that prompted President Kennedy to say in a press conference: "Life isn't fair." When we contemplate correcting for unequal impact of rising energy prices, the best we can do is to start from some arbitrary time base, without putting our stamp of approval on the situation in 1967, 1973, or any other year.

We lack a base period when burdens were shared equitably

Even then, conceptual and measurement problems remain. The question of whether income distribution has become more or less equitable is complicated. Higher transfer payments (Social Security, Aid to Families with Dependent Children, welfare, Medicare, food stamps, housing subsidies, and so forth) should be part of such an evaluation. Certain new disparities (not connected with energy) that have arisen since the base year raise the question of how prices relate to changes in real income (including transfers, changes in indebtedness, and the like). For some, lower real prices for food, clothing, and shelter offset price increases in energy; moreover, portions of those needs are not supplied out of current income. As we shall see later, energy outlays as a fraction of *income* tend to differ markedly from energy outlays as a fraction of *expenditures*.

Along with energy expenditures, income items may vary considerably by locality. And the likely new inequities by region, reflecting both energy supply and energy costs, involve perhaps as many complexities as the task of measuring fresh disadvantages for the poor in general.

Dollars-and-cents-type income is not the only criterion of a person's ability and willingness to bear burdens. Obvious examples are those aged who are poor in terms of current income but who may live comfortably and by conscious choice in well-insulated houses with paid-up mortgages, making few new purchases—so that energy-based inflation leaves them comparatively unscathed. That brings up the whole subject of changing asset values: rising energy prices tend to take a larger slice of income or savings, yet they also inflate the worth of existing holdings—such as conveniently located residential property. Even the values of human skills change, because of the shifting job market produced by a consciousness of the need to reduce energy intensity in the way we live, work, and play. Nor should psychological factors be ignored; there are undoubtedly some who could escape special regional energy problems if they wished, but who have personal reasons of preference that transcend the economic disadvantages.[4] They stay where they are.

The bottom line in this discussion is that, however defined, inequities exist today and they existed in 1973. There is some reason to believe that they will increase in certain respects as we move through the 1980s, as a more-or-less direct result of energy policies which were in the making in the late phases of the Carter administration and seem to have emerged as preferred in the 1980 elections. Thus, the distinct questions of social policy boil down to these: (1) Are we closer to or farther from an adequate provision of "equity" (however one interprets that) than we were in the early part of the past decade? And (2) are we satisfied as a nation with where we are now, or shall we take some specific steps in response to the energy price situation (cash payments, assistance in kind, rate reform, or more radical efforts) to provide more balance in burdens?

Divergent views on the need for action

The views of the conference participants, especially on the second question, diverged widely, and they undoubtedly reflect equally wide divergences in public opinion.

Herbert Stein, of the University of Virginia, commented on the possibility of creating a new Consumer Price Index, especially for very poor people. "If the study did reveal that poor peoples' CPI behaved significantly differently than

Dollars-and-cents-type income is not the only criterion

Inequities exist today, and they existed in 1973

the national CPI," he said, "I would want to see welfare programs indexed by poor peoples' CPI." But he added his own doubts that any great difference existed, and generally looked with skepticism at the case for energy-based income supplements: "It doesn't seem to me that energy, as a source or use of income, falls into a category that deserves special treatment."

Letitia Chambers, then staff director of the Senate Committee on Labor and Human Resources, ticked off several means of achieving equity as variously defined: (1) compensation to equalize expenditures on energy; (2) an adjustment which might equalize *percentages* of income associated with such expenditures; or (3) some more sweeping form of income redistribution through general welfare reform. But then she posed an even more basic question about equity that federal legislators must ask themselves: Does fairness demand that all states be uniform in the way they handle energy aid, or should any assistance system be related to individual needs? From the standpoint of political reality, her opinion was that need must be the guiding criterion—even though it would obviously be more difficult to satisfy in any administratively manageable manner. Moreover, the new administration's expressed intention to give greatly increased decision-making power to the states adds to that difficulty.

William A. Darity, Jr., an economist from the University of Texas, whose approach was avowedly Marxian, downgraded the whole discussion of income redistribution as secondary to the question of how society is directed. But, in what he characterized as a "second best" world, he favored focusing on specific inequalities; and affordable access to energy was high on his list of these. Darity suggested that emergency situations might arise in which quantity rationing would be needed, because time might not be available to await the normal workings of lump-sum transfers and market allocation.

Alperovitz was convinced that efforts in behalf of the poor had not kept up with the rising costs of energy (which he groups with food, shelter, and medical care as the primary necessities to which all citizens are entitled in some life-sustaining degree).[5] He cited calculations from the Department of Energy's Fuel Oil Marketing Advisory Committee, according to which the poor lost $14 billion in purchasing power in the three years 1978 to 1980 just through higher energy prices, and contrasted this with compensation programs on the order of only $4 billion.[6] When the real-income-depressing effects of slow national product growth and inflation are added to the calculation of average real income, he contends that the average family wage was maintained during the 1970s (to the extent that it was) only "by a very dramatic change in the participation rates for women"—more than a doubling of female jobholders in the course of a single generation. He concluded: "The average housewife has bailed out the average policymaker. Our failure in the economy has been repaired by women going to work."

The most common attitude, however, was probably the twofold observation voiced by Larry Ruff, then of Brookhaven National Laboratory. He began by saying, "The poor spend proportionately more of their income on energy, hence higher energy costs will hurt them more than proportionately. Hence, special programs are called for if this is regarded as a serious problem." But then he added, quickly, that "something should be done about that even if the data showed something different—even if the data were to show that the poor spent proportionately *less* of their income than the rich, or the more wealthy, on energy."

In other words, the problems of poverty (and, one might add, the problems of regional imbalances, sectoral differentials, and socioeconomic inequities between city dwellers and suburbanites or the rural population) can be addressed on their own. In the interest of achieving both social welfare goals and the objectives of our national energy policy, in fact, they should be.

Notes

1. *Reducing U.S. Oil Vulnerability: Energy Policy for the 1980s*, an analytical report to the Secretary of Energy prepared by the Assistant Secretary for Policy and Evaluation (Washington, D.C., GPO, Nov. 10, 1980).

2. U.S. Department of Energy, Energy Information Administration, *Monthly Energy Review* (March 1981) (Washington, D.C., DOE/EIA, 1981); and *Economic Report of the President, January 1981* (Washington, GPO).

3. In a provocative digression on the subject of less-developed countries, Alperovitz also mentioned the worldwide implications of any recession in the United States. Because of the gargantuan impact any slackening of U.S. trade may have on fragile LDC economies, Alperovitz proposed that safeguarding the health of our own system may be of more value to them than any amount of direct foreign aid that might be expected realistically within the next few years.

4. The same idea applies to those who strongly favor living in a detached house to apartment-dwelling, and to those who consider having a great deal of living space more important than other material comforts.

5. In connection with the fundamental point about rising energy costs versus rising income supplements, two participants provided information at the conference. Harold Beebout reported that Aid to Families with Dependent Children (AFDC) payments had risen much more slowly than the cost of living, while Social Security payments were supposed to be fully indexed to the CPI. Alan L. Cohen provided some specifics: for a family of four between 1975 and 1979, the weighted average of AFDC benefits plus food stamps lagged 7 percent behind the CPI. This national average masked even greater state-by-state differences, because the discrepancy was greater than this amount in thirty-seven states.

6. See chapter 5 (and especially pp. 68–72) for cost estimates presented to the conference and table 5-1 for data on federal energy assistance programs.

3

Do the poor get poorer?

No matter how one adjusts the data on the income, expenditure, or price front, precipitous rises in the dollar costs of energy during the 1970s hurt Americans at the bottom of the economic ladder more than they hurt those on the upper rungs. It was no surprise that fresh statistics presented at the RFF–Brookings Conference made that clear. As prices increase with recurrent market tightness, or with the introduction of synthetic and costlier-to-produce conventional fuels, poorer people will continue to be more vulnerable to new hardships. More of the same is in store, as U.S. energy policy moves past deregulation of petroleum toward decontrol of natural gas prices.

Why then is the subject controversial at all? Largely because the available data are inadequate and unreliable, and because terminology and concepts have been anything but precise. Until this situation improves, it will remain difficult to develop responses to the problem that can command widespread acceptance. This is troublesome, because another price shock could occur at almost any time and delay in developing policies could be costly. The solutions we hammer out are likely to be better ones if those who wage any new energy-and-welfare debate understand what the terms themselves mean as seemingly conflicting evidence is examined. Otherwise, those who begin from opposite perspectives might never overcome initial suspicions. Real issues will be fogged by bluster.

As has been said, a frequent source of discord is that people will use the same term to mean different things and different terms to mean the same thing. This is eminently true for the energy–equity debate.

A major cause of uncertainty and controversy lies in the substantial variations in reports from various authorities about how much energy is "costing the poor." In retrospect, it is easy to understand the apparent contradictions. As the discrepancies are tracked down, it becomes obvious that most are based on sincere (and easily reconcilable) differences in definition. No debate over energy policy and equity can get very far until each participant explains what he means when he says "energy" and what he means when he says "poor."[1]

We might start by asking how many poor people there are in the United States today and go on to consider what we mean when we say they "get poorer" for this or that specified reason. To the extent that we link this result with escalating energy costs, we also need to remember that "energy" itself is not homogeneous. Fuel prices have not risen uniformly, even in a given locality; and one form of energy is not always directly substitutable for another. Thus, energy expenses rise at quite different rates for people in diverse social, economic, and geographic circumstances, with different housing and transportation patterns and the often historical accident of using fuels in different proportions.

Energy expenses rise at different rates for people in diverse circumstances

Dispassionate analysis of this sort is less stirring emotionally than accounts, however accurate and upsetting, about the real-life threat of having some Americans freeze in the dark. Nevertheless, it is facts and their analysis, not emotions, that are the building blocks of a sensible national policy. And such analysis offers strong arguments for separating the country's treatment of its poverty problems from national pricing policies in regard to energy.

Defining the poor

First of all, who are "the poor?" Are we talking about nearly half our population (as is sometimes implied), or should we confine discussion to a much smaller fraction—those who are unarguably destitute? If our objective as a society is to help them, how do we go about finding them?

There are two principal yardsticks for identifying the poor, along with several minor ways. Far from being an exercise in bureaucratic mumbo jumbo, the distinctions have much to do with shaping one's evaluation of how affordable crucial energy services are. At the same time, it is not that one definition of poverty is correct and the others are all wrong. The point here is that a subject which is complex at best can become hopelessly garbled when people of good faith try to compare independent notes about different factors to which they validly assign the same name.

Perhaps the most widely recognized criterion is the "poverty index." Developed more than fifteen years ago by the Social Security Administration, it is essentially based on periodic computations of the nationally averaged cost for individuals or family members of obtaining a bare subsistence diet, derived from the 1955 food consumption survey of the U.S. Department of Agriculture. Under this formula, without considering local conditions or the specifics of any other expenses (such as clothing, shelter, utilities, or transportation), those family units whose cash income is less than three times this minimum food budget are generally labeled "poor."[2] On that basis using 1980 prices, a nonfarm family of four living in a high-cost area like Washington, D.C., would remain above the "poverty line" so long as its total cash income (before Social Security deductions) exceeded $8,450.

The Bureau of Labor Statistics offers a slightly more elaborate gauge—the so-called Lower Living Standard. This takes more of a family's normal living expenses into account, and it is adjusted by geographic area. While applying the poverty index suggests that there have been between 24 and 25 million "poor" Americans at any given time in recent years, the BLS standard yields a larger number—about 45 million persons in some 13 or 14 million households. To be specific, in the fall of 1980 the national average income needed to provide the Lower Living Standard for a family of four living in an urban metropolitan area was $14,044—nearly 170 percent of the poverty index for a household of that size. Finally, some programs make use of a third definition, those below an income that is 25 percent above the Poverty Line.

Poverty Line: $8,450
Lower Living Standard: $14,044

Of course, some people instinctively associate the poor with public assistance programs. In their minds, the category is restricted to those who are eligible for one welfare program or another—or perhaps only to households which actually apply for some type of public aid (for example, food stamps). Far from showing an unsympathetic attitude toward the economically disadvantaged, this method of categorization may correctly identify the "truly poor." Presumably, it excludes from consideration those persons who have substantial usable or spendable assets, even though cash income may be quite low during the year. But there are drawbacks to this definition too.

For example, some people simply refuse to accept public aid or are not reached by it, even though they could be considered poor by almost any test. A 1980 study by the Energy Department's Fuel Oil Marketing Advisory Committee (FOMAC) reported that only 40 to 50 percent of the low-income population was actually enrolled in public assistance programs.[3] Does this mean that the others are not poor? Eligibility for public aid can often be determined haphazardly by the states rather than in accordance with any nationally uniform formula. In fact, this has been the case with recent federal *energy* assistance programs, in which the respective states decide upon both eligibility and benefit level. And the fact that various states react in contrasting ways to the whole question of public assistance programs is suggested by the map reproduced as

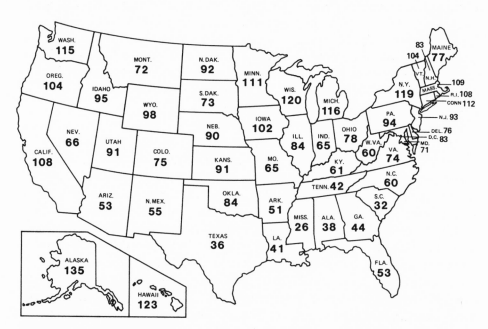

Figure 3-1. Average monthly payments per person (in dollars, as of June 1979) under the Aid to Families with Dependent Children program. Cost is now shared by the federal, state, and local governments, but the level of benefits (which is supposed to cover utility costs and all other basic needs) is set by the state. (Based on Sar A. Levitan, *Program in Aid of the Poor for the 1980s*, 4th ed., Johns Hopkins University Press, Baltimore, Md., 1980, p. 31.)

HIGH ENERGY COSTS: UNEVEN, UNFAIR, UNAVOIDABLE?

figure 3-1 on page 20, which shows the enormous variation in average payments under the program of Aid to Families with Dependent Children (AFDC). In Texas these payments are $36 per person and in Mississippi only $26, yet fully one-quarter of the states peg them at more than $100 per month. Those discrepancies are too great to explain on the basis of differences in the cost of living alone.

Income levels and energy expenditures: A complex relationship

There are many purposes in raising these matters here, and the reasons should become clear as we move into a discussion of options for energy–equity policies later on. For the moment it suffices to point out the obvious fact that—even though statistics on the subject may diverge considerably because "poor" is being defined in different ways—poverty is related to income. And, because we need to see how the effects of changing energy prices relate to income, a logical next step is to compare the real changes over time with the average income for U.S. households (see figure 3-2).

This graph shows some interesting and important trends:

1. Over a couple of decades, the real price of energy to end users has not, all in all, risen much; the unit cost of electricity has even dropped. But the increases have come in two big jolts, 1974 and 1979—just as household income was faltering. That is what has hurt! And the real price of heating oil has nearly doubled.

2. On the basis of energy content, natural gas has been and continues to be a relative bargain; and its average price nationwide remained remarkably steady until very recently. (Coal would look even cheaper if priced on the basis of contained energy, but is omitted because it is rarely used as a household fuel.)

In a country as large and diverse as the United States, however, national averages conceal substantial price variations. The geographic breakdown of energy prices summarized in table 3-1 illustrates the point. Because we will first address residential energy use, it includes liquefied petroleum gas (LPG), or "bottled gas."

The purpose of including such a table at this point is not principally to note that residents of the Northeast must pay higher-than-average prices for all fuels

National averages conceal substantial price variations

while those who live in the West are favored by lower-than-average prices. That happens to be true, but it is the topic to be addressed in the next chapter. Here we

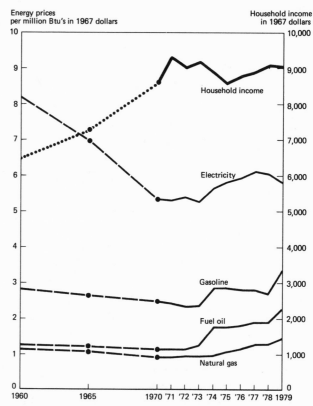

Figure 3-2. Energy prices and household income in constant dollars, 1960–79. Household income derived from *Current Population Reports* Series P-60; price data through 1978 derived from U.S. Department of Energy, Energy Information Administration, *State Energy Fuel Prices by Major Economic Sector, 1960 through 1977;* 1979 derived on percentage change from 1978, *Monthly Energy Review* (November 1980) DOE/EIA-0035 (80/11).

HIGH ENERGY COSTS: UNEVEN, UNFAIR, UNAVOIDABLE?

Table 3–1. Average Prices of Residential Energy Sources Delivered to User—Nationwide and by Census Region, April 1978–March 1979
(dollars per million Btus)

Region	Electricity	Natural gas	Fuel oil and kerosene	Liquefied petroleum gas
Nationwide	$12.10	$2.74	$3.93	$5.09
Northeast	15.34	3.42	3.98	7.93
North Central	13.64	2.57	3.82	4.55
South	11.75	2.85	3.94	5.16
West	8.28	2.30	3.77	4.18

Note: For states included in each region, see figure 4-1.

Source: Residential Energy Consumption Survey: Consumption and Expenditures, April 1978 through March 1979, DOE/EIA-0207/5 (Washington, D.C., DOE/EIA, July 1980) table 3.

are focusing on price impacts *on the poor*, and the intent is to show how traditional choices among more or less substitutable energy forms may vary those impacts enormously in different areas, when regional price differences are taken into account.

Because of the costs of long-distance pipeline delivery from the major domestic producing areas, natural gas (which is generally cheap) is less of a relative bargain in the Northeast—despite price control. It is easy to see why the poor in that area have suffered a greater shock from rising oil prices than the poor in other regions—where there had been an earlier economic incentive to favor gas over oil when it was available.[4] Thus, the Northeast was doubly disadvantaged: less able to use the nationally less expensive energy, and finding fuel oil, the local choice, rising disproportionately in price. Similarly, the special price position of electricity in the West (a long-time beneficiary of great hydroelectric development) has skewed the energy mix there from the national norm.

Energy sources and uses

By themselves, however, the relative prices of various energy forms do not fully dictate the energy mix in a given home either. Matching form to intended use is a major determinant; and sometimes availability and even accident play a role. Figure 3-3 shows in different ways how U.S. households utilize different forms of energy.

According to these data, almost all U.S. homes and apartments are equipped with electricity, and nearly two-thirds have piped gas. Only one in five now uses heating oil, and the use of LPG (bottled gas) is quite small. In terms of energy content, natural gas is by far the most important residential source—surpassing all others combined. But analyzing direct *expenditures* on energy yields a different breakdown: U.S. householders spend less on natural gas than physical use data

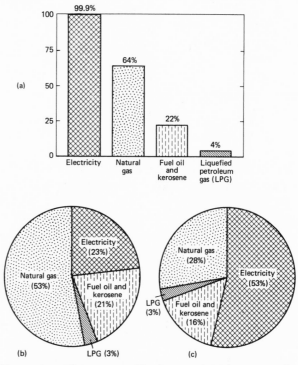

Figure 3-3. Three ways of looking at the relative significance of energy sources common to U.S. households, 1978–79.(a) Frequency of use, (b) amount of use (in British thermal units), and (c) dollar cost for use (based on table A-1).

24

would lead one to suppose. The explanation lies, of course, in the relatively low price of natural gas.[5] Electricity appears to account for an inordinate share of

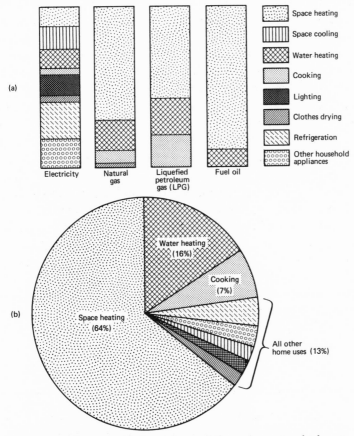

Figure 3-4. A nationwide breakdown of the end uses to which residential energy is put. The bar graphs indicate the uses to which individual energy sources are applied in the home. The pie chart shows the overall allocation of all home energy sources combined. Note that figures are for 1975 consumption. (Based on data from Jill G. King, *Residential Energy Consumption by Functional End Use in 1975*, Mathematica Policy Research, Washington, D.C., 1979).

household energy expenses—but the comparison necessitates at least one further bit of analysis (see figure 3-4).

The various forms of household energy are only partly interchangeable. More than half of all residential electricity is used for applications in which other energy sources are either not practical at all or have captured only a small share of the market (lighting, refrigeration, space cooling, and various appliances). This is one reason why consumers have been willing to pay the traditionally higher costs per Btu of electricity (see figure 3-2). Another is that measured at the point of use, electricity is extremely efficient in what it does, while home heating fuels may suffer significant end-use losses. By the same token, electricity has an "embodied energy cost" of its own—the 70 percent or so of energy lost in the process of generating and delivering electricity. One might assume that any competitive appeal it has in conventional home-heating applications rests largely on convenience and cleanliness. In terms of cost, it means that the losses between energy content of the primary source and energy availability at point of use are taken care of for the most part by the time electricity reaches the consumer. Not so for oil and gas. In any event, as shown in figure 3-4, those nonheating applications represent a relatively small percentage of the total energy used in American homes. Four-fifths of all residential energy goes into providing either space heat or hot water. These happen to be the only energy services which oil products provide; so for those people who use heating oil as their energy source, its adequate supply translates into "keeping warm." Almost the same is true for natural gas, although cooking and drying of clothes account for 8 percent and 3 percent, respectively, of the gas consumed in gas-using homes.

The various forms of energy are only partly interchangeable

How households allocate their energy purchases

So far, we have looked only at all U.S. households and tried to portray the complexity of the energy use pattern. However, all of the foregoing is invaluable in interpreting figure 3-5, which takes us back to the equity problem. It shows how much households at various income levels spent on electricity, natural gas, and fuel oil during 1978–79—just before the last big spurt in oil prices.

The information here comes from the National Interim Energy Consumption Survey (NIECS), which examined 4,000 households in the continental United

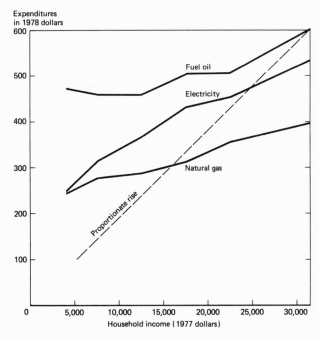

Figure 3-5. Relationship between income and direct household expenditures on various energy sources. (From the conference paper by Harold Beebout, Gerald Peabody, and Pat Doyle, with data for this figure based on U.S. Department of Energy, Energy Information Administration, The National Interim Energy Consumption Survey.)

States, starting in 1978. This survey was intended as a pilot program for the far more comprehensive Residential Energy Consumption Survey (RECS), which got under way in late 1980. Unlike some earlier studies, NIECS did not risk so many of the errors inherent in asking those surveyed to estimate what their energy consumption and expenditures were; in most cases, billings were checked directly with the utilities and fuel companies involved. Because the survey began before the end of 1978, expenditures were correlated with household incomes during the most recent year for which they could be reported in full—1977.

The dashed line in figure 3-5 shows the slope at which direct household energy expenditures would rise if those costs represented the same percentage of each household's income.[6] Of course, they do *not*. We might have surmised this intuitively; but here the fact is displayed graphically.

In the case of each of the three energy sources, the direct outlay for energy rises more slowly than income as we climb the economic ladder, and a smaller and smaller percentage of income is directed to acquiring energy.[7] But the figure shows a little more. Up to an income of about $15,000, expenditures on electricity roughly parallel the slope of "constant proportion." This suggests that—even though expenditures for electricity eventually decline relative to income as it rises over the whole range—there is considerable scope to add to the stock of electricity-operated conveniences until some kind of "saturation level" is reached. At the lower end of the income range, those services tend to be added rapidly.

In the cases of the two fuels used primarily for heating, however, the situation is quite different; the entire slopes, while not flat, exhibit only a moderate increase—especially up to an income of $15,000. Whether the survey was dealing with families whose annual income was $5,000 or six times that, it found that the average expenditures on natural gas and fuel oil were not nearly as far apart proportionally as incomes were. This suggests that energy required for basic comfort levels differs little by size of income and that such levels constitute a high priority for all, regardless of income. This is especially intriguing in view of the natural assumption that families with higher income can afford more living space, and thus have larger areas to heat. Reality, no doubt, is more complex. For example, one explanation is that lower-income families tend to live in older buildings—which are likely to lack storm windows, storm doors, and insulation, and are, in general, not energy-efficient. The NIECS data confirmed earlier, more geographically limited findings to this effect. It is also probably reasonable to assume that heating systems in such structures do not operate as efficiently on the average as those in higher-cost housing. And there may be other factors, such as sheer inertia of older people continuing to live in the homes they have lived in for many years.

The direct outlay for energy rises more slowly than income as we climb the economic ladder

Expenditures on natural gas and fuel oil were not nearly as far apart proportionally as incomes

HIGH ENERGY COSTS: UNEVEN, UNFAIR, UNAVOIDABLE?

As a matter of fact, a further breakdown of the NIECS data shows other variables which also seem to relate to differences in direct residential energy costs. Households including five persons or more spent nearly twice as much on natural gas as did single-person households ($403 per year, compared with $221). The larger households also spent considerably more on electricity ($536, compared with $227). Yet this "logical" pattern falls apart in relation to fuel oil expenditures, where the average outlays by households consisting of five or more persons ($591) were only about 30 percent greater than the $457 spent by solitary householders on that fuel, and three-person households spent less than two-person households. These and other anomalies listed in table 3-2 spur one to look elsewhere for possible explanations. Geography and climate are two that will be considered in the next section.

Meanwhile, we will consider two other general aspects of the rising energy cost-burden for those with the lowest incomes. The first is the cost of gasoline, and the second is the cost paid for the energy "embodied" in almost all goods and services anybody purchases (that is, the indirect purchase of energy).

The case of gasoline

The importance of gasoline expenses is suggested by table 3-3, based on the 1980 FOMAC Report cited above. It suggests that low-income families spent half as much again on transport energy as on household energy. If anything, these statistics may understate the significance of automotive fuel to many poor people, because the numbers quoted are average expenditures, and approximately half of all the households in the lowest income categories do not own automobiles. This means that some poor persons who do drive their own cars might easily have been spending more than one-third of their total cash incomes in 1979 on the combination of household energy and gasoline.

Low-income families spend half as much again on transport energy as on household energy

As in the case of fuel oil, price increases on gasoline undoubtedly were noticed more because they came in two distinct jumps, rather than building up gradually (again, see figure 3-2). Furthermore, those who were driving older (less mileage-efficient) cars and could not afford to replace them on short order were doubly penalized by their economic circumstances.

A question that is raised frequently concerns the extent to which expenditures on gasoline (or replacement transportation) are discretionary and

Table 3–2. Average Expenditures as a Proportion of Income for All Fuels, and Average Expenditures for Specific Fuels by Various Socioeconomic Characteristics of Households, April 1978–March 1979

Socioeconomic characteristics	All fuels Average expenditures (dollars)	All fuels Expenditure as percentage of income	Type of fuel (average dollar expenditure by users) Natural gas	Type of fuel (average dollar expenditure by users) Fuel oil and kerosene	Type of fuel (average dollar expenditure by users) Electricity	Type of fuel (average dollar expenditure by users) Liquefied petroleum gas
Total households	724	8	314	501	390	241
1977 household income						
$3000 to $4999	522	14	244	470	246	193
$5000 to $9999	627	8	278	457	311	226
$10,000 to $14,999	659	5	287	456	365	235
$15,000 to $19,999	769	4	312	501	430	313
$20,000 to $24,999	816	4	354	504	453	244
$25,000 or more	938	3	398	602	535	285
Poor households [a]	582	[b]	279	486	274	190
Age of household head						
29 or less	603	7	272	451	330	211
30 to 44	809	6	331	531	449	292
45 to 59	816	6	351	517	447	256
60 and over	644	13	294	478	324	194
Household size						
1 person	489	10	221	457	227	220
2 persons	669	7	293	494	362	199
3 persons	776	7	351	479	425	269
4 persons	853	7	353	506	476	255
5 or more	959	8	403	591	536	321
Household composition						
With children						
Female head	748	13	336	542	380	211
Male head	871	6	371	525	488	295
Without children						
Female head	507	11	234	474	239	218
Male head	649	7	281	475	349	201

(Continued)

Table 3–2 *(continued)*.

Socioeconomic characteristics	All fuels		Type of fuel (average dollar expenditure by users)			
	Average expenditures (dollars)	Expenditure as percentage of income	Natural gas	Fuel oil and kerosene	Electricity	Liquefied petroleum gas
Employment status						
Head married	796	6	342	506	442	247
Head or spouse employed full time	847	5	358	533	468	253
Both spouses employed full time	793	4	330	494	457	290
Neither employed full time	666	10	320	440	352	174
Head not married	577	11	261	490	234	224
Employed full time	598	7	269	517	304	254
Not employed full time	558	15	253	469	267	206

Source: Reproduced, with minor modifications, from the conference paper, "The Distribution of Household Energy Expenditures and the Impact of High Prices," prepared by Harold Beebout, Gerald Peabody, and Pat Doyle; and based on the National Interim Energy Consumption Survey.

[a] Defined as households having family incomes below 100 percent of the official "poverty line."

[b] The available income data for this group are insufficient to compute this ratio.

to what extent they fall into an "essential" category. Is there not some auto travel whose cost would be as difficult to avoid as the costs of minimum home heating? There are no good universal answers, but some data help illuminate the topic. On a nationwide basis, it was estimated recently that commuting to work accounts for about one-third of all vehicle-miles traveled in household autos.[8] On the other hand, the relatively large difference in transportation-gasoline expenses between low-income families and those of median income, shown in table 3-3, hints strongly that extended auto use (that is, beyond the seventeen- or eighteen-mile trip to work and back which is the U.S. average[9]) may be considered discretionary in that it is related to what families can afford. However, residence–workplace distances could also be involved in these larger differences. The distinction between "essential" and "discretionary" is not, of course, confined to driving. It applies equally to household uses, but is most frequently brought up with regard to transportation.

More precise and more up-to-date information about gasoline consumption and costs should be available soon. Driving habits were recorded by NIECS, but results of that part of the survey were published too late to be considered here.

Table 3–3. Estimated Direct Energy Costs per Household for Residential Uses Plus Transportation, Showing Comparison Between Low-Income and Median-Income (Nationwide Averages)—1978–79
(absolute figures in dollars)

Income category and year	Income	Household energy expense	Transportation/ gasoline expense	Total direct energy cost	Percentage of income spent on energy
Low income [a]					
1978	3,401	604	228	832	24.5
1979	3,549	699	350	1,049	29.5
U.S. median income					
1978	17,640	768	910	1,678	9.5
1979	18,875	889	1,227	2,116	11.2

Source: U.S. Department of Energy, Fuel Oil Marketing Advisory Committee, *Low-Income Energy Assistance Programs: A Profile of Need and Policy Options* (Washington, D.C., DOE, July 1980) tables I and II, pp. 9–10.

[a] The income used here is the median within income categories which do not correspond exactly to those shown in table 3-4. Another reason for discrepancies between the two tables is that this table is based on projections of 1975 consumption patterns, while table 3-4 reflects actual billings for a more recent period.

HIGH ENERGY COSTS: UNEVEN, UNFAIR, UNAVOIDABLE?

Buying energy indirectly

The problem of being compelled to project from old data also appears when we try to estimate the economic impact that rising energy prices produce at various income levels because of the increased costs of manufacturing, packaging, and delivery. Analyses of "embodied" energy cited at the conference used updated energy prices,[10] but still depended on 1972–73 consumer expenditure survey and input–output data. Buying patterns have changed over nearly a decade, in some cases as a direct response to changes in energy prices, but also because of many other social, economic, and technical developments. Furthermore, we know that industry became generally more energy-efficient during the 1970s.[11] Having expressed this caution, however, we have little choice but to assume, perhaps not unreasonably, that certain underlying trends along the poverty-to-affluence scale continue.

Some findings, made at specific earlier dates, do exist. They suggest that the ratio between the calculated cost of embodied energy (that is, *indirect* energy outlays) and a household's *direct* expenditures on energy appear to rise as overall money income and expenditures increase.[12] One analysis published in 1976 shows that, shortly after the Arab Oil Embargo of 1973–74, a family whose income was in the very lowest ranks (averaging only $1,270 per year) was estimated to have spent 63 percent as much on embodied energy as it did on direct payments for fuels and electricity. A household in a much higher income category (with an average income before taxes of just over $30,000) spent 85 percent as much as its direct energy payments on this additional, "hidden" energy cost. At an intermediate level (around $11,000 in income), the ratio was likewise intermediate—76 percent.[13]

But while they rise relative to direct energy costs, as income rises, even these "indirect" costs are at least mildly regressive in themselves. In the lowest income category, embodied energy accounted for 8.4 percent of all household expenditures. As the income brackets rose, the percentage dropped to 7 and 6.8 percent, respectively, for the levels cited. Thus, the addition of embodied energy costs merely tends to moderate the sharply regressive impact of rising energy prices.

The burden of rising energy prices: A summary

The full significance of energy prices within the budgets of those with the lowest incomes seems to be very great. If we combine the recent data on direct energy expenditures from table 3-3 with the best available calculations on the cost of embodied energy, we come to the startling approximation that—as a national average—the poorest U.S. families in 1979 were required to spend nearly half of their reported income to pay for the direct and indirect costs of energy. Yet two qualifications need to be made.

First, such regressive impact is much greater when energy expenditures are related to household *income* than when they are matched against household *expenditures*. This is because, as table 3-4 shows, the average spending of those in the lowest income categories greatly exceeds their monetary income. Such situations are by no means rare; similar averages apply to income groups that make up about one-quarter of the population. Undoubtedly, a large percentage of people in these categories receive public assistance—either in kind (so that it does not show up either as income or as expenditure) or in cash (such as Social Security and other assistance payments which, the Census Bureau states, are typically underreported in household surveys). But there are various explanations for the rest. For instance, many are probably retired persons who draw on savings and other assets for current expenses. Some others may be temporarily unemployed, still others are workers who have returned to school on a short-term basis to complete specific courses of study, and some are recent widows—who have not lately been part of the full-time labor force, but who will eventually return to it. Finally, there is the highly controversial issue of the magnitude of the "underground" economy—with its *unreported* income. It is difficult to judge how many with low recorded cash income (and which ones) should actually be counted among the "poor" when we try to measure the impact of rising energy prices on those who live in poverty. The best we can say is that using income as the measure of comparison undoubtedly exaggerates the gravity of the true situation. But how great an exaggeration only future research will reveal.

Second, the definition of "poverty level" is outdated and inadequate. Yet it is so convenient that both politicians and academicians have tried to hang onto it

Average spending of those in the lower-income categories greatly exceeds their income

Table 3–4. Direct Energy Expenditures, Averaged for Households in Selected Income Categories
(absolute figures in December 1974 dollars)

Income group	Low	Medium	High
Average money income before taxes	1,270	10,907	30,100
Average consumption expenditures	2,818	9,769	21,028
Direct energy expenditures	377	898	1,674
Percentage of income devoted to direct energy expenditures	29.6	8.2	5.6
Percentage of average consumption expenditures devoted to direct energy	13.4	9.2	8.0

Note: The "low" group is represented by the lowest 4 percent in the income distribution. The "medium" group is represented by the 13 percent located just below the median income, and the high group is represented by the 7 percent just below the top 2 percent of the income range.
Source: James P. Stucker, "The Impact of Energy Prices on Households: An Illustration," Paper P-5585 (Santa Monica, Calif., The Rand Corporation, January 1976).

by using some multiplicative factor; for example, some would extend eligibility for various aid programs to those with 125 percent (or even 200 percent) of a poverty level income. But the discomforting reality is that national averages offer at best a vague statistical measure of the poverty problem and the energy-price problem. They give no key at all to the appropriate solutions, in terms of the statistics' real counterpart.

One commonsense observation seems to be indisputable. High prices in themselves are less difficult to cope with than sudden upward *changes* in price. Over time, people at any economic level adjust to high prices for a product or service by modifying their consumption habits accordingly. As it is more difficult for the truly poor to do so, a farsighted social welfare policy might well concentrate on speeding up those adaptations rather than pretending that the price adjustment can be postponed indefinitely. Some ways of doing this are suggested in chapters 5 and 6.

Sweeping all other complications aside, especially those that have to do with the difficulty of identifying the real-world counterparts of the statistical entries, have the poor lost ground as a result of escalating energy prices? Of course they have, but so have most Americans. The opportunity for saving has been reduced. Compensation by conserving is not always possible and not necessarily painless. Choices in life-styles are somewhat more limited.

In general, have the poor suffered more than the rest of us? The easy, overall answer is, again, yes. Anybody clinging to the edge of economic survival is vulnerable to even a slight disturbance. There are fewer escape valves, and using them is more painful: to consume less of the more costly commodity or service, to consume less of other goods and services, to rely more heavily on public assistance, or to seek additional employment if possible. Yet, there are surely many individual exceptions too. The degree of economic penalty to date has depended heavily on what a person's or a family's energy habits were to begin with (including where and how they lived) and what tradeoffs were available. It has also depended on one's role in the economy, rather than on income alone. For example, the energy-related effects on the U.S. auto industry have been felt all along the socioeconomic scale.

What ways do we have to assess the economic penalty of rising energy prices? Knowing the absolute dollar cost of higher energy prices to an individual or a family (both in direct fuel outlays and in "embodied" costs) does not in itself give us a useful measure. Calculating the cost as a percentage of real income is a somewhat better gauge. Determining its percentage relationship to expenditures is probably even better. And it would probably be best of all if we could calculate those specific costs as a percentage of each household's disposable income, taking into account all assets, receipts in kind, and a host of intangible factors as well. As a practical matter, there is no way we can do that. Policy must work through *categories* defined by specific criteria. These will be imprecise at the margin, omitting some that should be included and vice versa.

Compensation by conserving is not always possible and not necessarily painless

But trying to determine which *individuals* have been hurt inordinately by rising energy prices (and thus might qualify for aid) would involve insuperable complexities.

Because it is so difficult to disentangle the relationship of energy use to income and expenditure status, divorcing energy-price policy from the question of how and how much the government at various levels should help the neediest among us appears attractive. It is easier both to judge whether someone is poor, by some convenient definition, and to determine the level of general assistance that would relieve hardship to some established level than to do so with specific reference to the role that energy plays in the poverty syndrome. Whether this is the way to go in practice we shall discuss at the conclusion of this report.

Notes

1. Note that placing the word poor in quotation marks does not indicate skepticism or unreality, but only that we are using the term in a technical, definitional context.

2. There are a few modifying elements to the formula, but they are applied uniformly on a nationwide basis. For example, the prescribed "subsistence level" is automatically lower for farm families, elderly persons, and households containing no adult male. There is a useful discussion about the problems of measuring poverty in the opening chapter of Sar A. Levitan's *Programs in Aid of the Poor for the 1980s* (4th ed., Baltimore, Md., Johns Hopkins University Press, 1980).

3. *Low-Income Energy Assistance Programs: A Profile of Need and Policy Options*, the second report issued by the Fuel Oil Marketing Advisory Committee of the U.S. Department of Energy (FOMAC), July 1980, p. 41. Subsequent references to this document identify it as the FOMAC Report—1980.

4. Data presented at the conference by Harold Beebout, Gerald Peabody, and Pat Doyle showed that more than half of all houses in the Northeast were still using fuel oil as their primary heating fuel in 1978—three times as many as in the South, which was the runner-up region in reliance on fuel oil. Between 1970 and 1978, all regions came to depend less on fuel oil for heating. The major shift in the North Central region was to gas, which rose from 67 to 74 percent in the percentage of homes it heated. For the other two regions, the big gainer was electricity, which went from 13 to 27 percent in the South and from 12 to 20 percent in the West.

5. This portends a widespread fresh shock if the decontrol of natural gas prices (now scheduled to take place by 1985) is speeded up—a step which is highly desirable from the standpoint of national energy policy because it would encourage domestic production (as opposed to mere exploration) and discourage less-than-optimal use of natural gas which may take place now as a result of the fuel's controlled, low price.

6. The "line of constant proportion" was suggested by Mark N. Cooper, director of research for the Consumer Energy Council of America, a conference participant.

7. Table 3-2 (on page 30) suggests how great the disparity may be. For those with incomes between $3,000 and $5,000, expenditures on all forms of household energy during a twelve-month period that included the 1978–79 winter were equal to 14 percent of reported income. For those with incomes above $15,000, the average percentage dropped to 4 percent or less.

8. J. Hayden Boyd, "The Microeconomics of Response to an Oil Import Disruption." Paper presented at the Second Annual North American Conference of the International Association of Energy Economists, Washington, D.C., October 6, 1980, table 12, p. 38. Boyd's calculation of 31.1 percent was based on the 1978 National Transportation Survey.

9. Ibid.

10. Robert Herendeen and Charlotte Ford, "Energy Cost of Living, 1972–73," Document No. 311, Energy Research Group, Office of the Vice Chancellor for Research, University of Illinois (1980); and James P. Stucker, "The Impact of Energy Prices on Households: An Illustration," Paper P-5585 (Santa Monica, Calif., The Rand Corporation, January 1976).

11. Between 1973 and 1980, U.S. industrial output rose by 13.3 percent while aggregate energy consumption by the industrial sector increased by a mere 2.8 percent. Still, it is difficult to guess the changes in energy intensity for the production of individual commodities without thorough study.

A thought-provoking example is the typical personal automobile. About ten years ago, R. Stephen Berry computed the energy input in its manufacture at the equivalent of approximately 1,250 gallons of gasoline (in "Recycling, Thermodynamics and Environmental Thrift," *Bulletin of Atomic Scientists* (May) 1972). Since then, the average weight of U.S. cars has dropped, but their manufacture now is likely to involve a higher percentage of energy-intensive materials—such as aluminum- and petroleum-based plastics. To complicate matters, high interest rates and recessionary pressures may encourage American car-owners currently to hold onto automobiles longer so that the initial energy investment should be prorated over a longer period of operation. Lacking more current analysis, it is problematical whether the "embodied energy cost" in owning a car is higher or lower now when measured in British thermal units. Nevertheless—considering a geometric rise in energy prices (even including that of a manufacturing fuel such as coal)—there is no doubt that the dollar outlay for the embodied energy has risen.

12. Note that here we are comparing expenditures for two different kinds of energy aggregates, not energy expenditures and incomes.

13. Calculations based by Harold Beebout on an analysis by Stucker, "The Impact of Energy Price Increases on Households."

4

It depends on where you live

Equality is an integral part of our American political creed. Equity is a goal that U.S. legislators stress continually. Yet we pride ourselves also on diversity within this continent-sized country of ours, and nothing reveals the dissimilarities in opportunity or burden among us more vividly than our nation's political geography. It has always produced sharp differences we were willing to accept (even flaunt), so long as the balances did not shift too suddenly, and so long as the tide rose fast enough to raise all the ships.

We are either city dwellers or suburbanites or rural residents—with distinct advantages and disadvantages accruing to each of us from this. Some of our home areas are sparsely populated, while others are packed with people. Our regional climates may be cold, hot, or gentle. We also live in partly autonomous states which (like sovereign countries) may be net importers or net exporters of energy in various forms. It is only natural that a major economic development such as the changes of the past few years in energy prices and supply conditions would produce differential impacts based on where one lives. What we need to explore is whether the various favorable and unfavorable changes tend to cancel each other, or whether the imperfect geographic equity which previously satisfied us—or did not trouble us deeply enough to become a major issue—is undergoing such trauma that some new national adjustment will be required.

The first official report of the 1980 Census brought wails from areas that had lost population, because apportionment of political representation (and, to some extent, distribution of government funds) is played as a "zero-sum game." If somebody wins, others must lose. In this case, the old North and the big cities lost. The winners were the South, the West, and suburban or rural areas.

In mid-January 1981, a report issued by President Carter's Commission for a National Agenda for the Eighties caused a further outcry when it recommended (with some dissents) that the federal government should stop efforts to slow or reverse these shifts in population, business, and industry.[1] It urged that programs aimed at sustaining the economic health of *places* be retargeted toward *people*—wherever they are. And it suggested that government encourage mobility toward whichever localities were already offering the greatest economic appeal. To our knowledge at least one state (Illinois) has begun to consider, though has not as yet instituted, "resettlement bonuses" to encourage its jobless residents to move someplace else— where economic prospects are brighter, and Marquette County (Michigan) pays bus fares for those who want to seek jobs out of state.

As yet, the public debate over this proposed element in an Agenda for the Eighties has not been correlated specifically with the continuing discussions of our national energy requirements; but at some point the two are bound to overlap. When that occurs, those who study the question will find the interrelationships even more complex than those involved in the impact of rising energy prices on the poor.

It is reasonable to speculate that energy price rises since the early 1970s have contributed to the southward and westward movements; if so, however, one should not assume that this will continue to be the case. As for cities, they should find that past and future increases in energy prices will reward and encourage in-town living—although a full-fledged reversal of flight to the suburbs cannot be expected at any early date.

We start by examining the factors in large-scale regional shifts, then look at urban versus nonurban effects, and finally speculate a bit about the future.

Energy-rich and energy-poor states

A recent study by Chase Econometrics provides a convenient starting point by rating fifty states according to their relative energy supply and demand status in 1976.[2] When energy production is divided by energy consumption in each case (and 100 percent is the expression used to signify equivalence between production and consumption), Louisiana and Wyoming scored far above the rest—with 406 and 405 percent, respectively. At the other end of the range were Delaware, Hawaii, and Rhode Island—essentially nonproducers of any source of energy. The results of the Chase study are presented in figure 4-1. Coding a U.S. map shows how the location of energy-surplus or reasonably energy-self-sufficient states relates to the four principal census regions—to which we referred in the preceding section and which we will mention again.

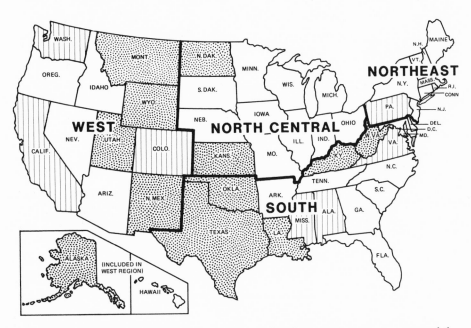

Figure 4-1. Relative self-sufficiency of states within four major census regions of the United States, as of 1976. *Dotted areas* indicate states producing mineral energy resources equal to or greater than their own composite consumption of all Energy. *Vertical lines* indicate states whose production of such resources represents at least half of their own composite consumption. *White areas* indicate states in which energy production is less than half of composite consumption. Note that the production data omit hydro- and nuclear-based electricity. (From Bureau of the Census and Chase Econometrics, as reported in the *National Journal*, March 22, 1980.)

Only two of the dozen states with "as much energy as they need" lie outside the South or West, and one of these twelve (Kansas) adjoins both regions. Of the six other states that produce energy corresponding to at least half of their own requirements, three are western and two are southern. In short, the South and West are the "haves" on the U.S. energy map.

In the discussion that follows we lean heavily upon the presentation made for the conference by William H. Miernyk of West Virginia University. To demonstrate the economic significance of this pattern he falls back upon a tool normally used only in discussion of international trade: "terms of trade." Because interstate transactions are not recorded in the same fashion as trade between nations, one can only simulate such "terms"; but the analysis can be made more credible by considering only commodities that are relatively homogeneous and only products for which the states in question are substantial contributors to national output. When putting together the relevant data, one may quite realistically describe what would have happened between 1970 and 1978 if energy-poor states had been required to barter their own goods for fuel from one or more of the energy exporters within our own national borders.

For example, processed tomatoes (a significant product of New Jersey) lost more than three-fourths of their "buying power" in terms of Texas natural gas, nearly two-thirds of their exchange equivalent in terms of West Virginia coal, and about three-fifths of what they had been worth in a "swap" for Oklahoma crude oil. Massachusetts cranberries, Minnesota barley, and Illinois corn suffered similar deterioration in exchange value.

A single exception is worth noting, if only because of its very rarity. Illinois hogs appreciated slightly during the period (about 5 percent) when compared with Oklahoma crude. And it should be added that some states in the South and West would *also* have been losers in trade between 1970 and 1978: for instance, Georgia (a mere 6 percent self-sufficient on the basis of its own energy production) saw the relative value of its broilers drop by 40 to 70 percent.

According to Miernyk (whose calculations are displayed more fully in table A-2), the marked shift in favor of the three basic energy forms was unequivocal—so transfers of wealth from "energy-poor" states to "energy-rich" states would be a fair assumption. Relatively more spending in the latter could stimulate local business, providing new job opportunities that might encourage immigration from areas which at the same moment were experiencing difficult times because of a suddenly less favorable balance of domestic trade.[3]

Employment and population shifts

Next, he moved along to enterprises whose products might be more difficult to compare in this way, striving to determine whether such businesses themselves had been lured into relocating or transferring employees to areas where energy was relatively more plentiful (and cheaper). He proposed an analysis of effects on various industrial and commercial activities by dividing such activities into three categories—according to the manner in which we might expect them to respond to rising energy costs:

1. *Sheltered activities.* This highly heterogeneous category is actually the largest. Besides those whose energy requirements are minimal, it consists of all activities which are more or less bound to their present locations—for example, because of the importance of local resources (such as primary fuels or minerals) or transportation facilities. It also includes those whose products or services have no serious "outside" competition within a locally limited market (which often means that higher energy costs can be passed along immediately and entirely to the consumer, even if the activity is energy-intensive). Some of their customers may be susceptible to relocation appeals; and the lower demand which results from higher prices (in all but the most inelastic situations) might diminish output. Yet these "sheltered" enterprises are unlikely to move.

2. *Directly vulnerable activities.* These specialize in energy-intensive products or services which must face national or broad regional competition. One generic example includes plants involved in chemical production. Local rises in energy prices could persuade them to reduce output or relocate.

3. *Indirectly vulnerable activities.* These have as their major customers one or more directly vulnerable enterprises whose disappearance would affect them gravely. An example might be a rail line associated heavily with an energy-intensive chemical complex. Energy price increases are likely to

affect them adversely, especially in relation to similar enterprises in areas that have not been hit as hard.

Miernyk analyzes U.S. industry in two years—1970 and 1977; and he subdivides employment statistics for the country among the ten Standard Federal Regions (SFRs) instead of the four fairly gross Census Regions we use here for the sake of simplicity. Perhaps the most striking fact is not directly related to interregional shifts at all; it is that overall employment in "directly vulnerable" activities dipped during the seven-year span (from 20.7 to 18.1 percent of all U.S. employment) while the share of jobs in "sheltered activities" was expanding nationwide (from 74.5 to 77.6 percent).[4] This might be interpreted as a universal shying away from those enterprises which have been hurt most by increasing energy costs and which might anticipate a further drubbing in the future, but the categories are too gross and the factors involved too numerous to draw any firm conclusion as to primary causes.

In regard to interregional changes, using the finer geographic breakdown of the SFRs gets fairly close to "energy-poor versus energy-rich" comparisons; but the data show less than an overwhelming trend. There has been some employment shift of directly vulnerable activities from the "have not" areas to the "haves," with the two energy-richest SFRs each showing a pickup of about 25 percent,[5] while energy-poor New England was showing an undeniable decrease. Yet some changes defy explanation by this theory. The record of one five-state region in the mid-Atlantic area was even worse than that of upper New England, even though the former included a sizable "energy exporter" (West Virginia—which produces 232 percent of its own energy consumption equivalent) and two more states that are in fair energy balance (Virginia, 89 percent; and Pennsylvania, 63 percent). Miernyk is thus justifiably cautious in labeling native energy resources as only *one* factor (and not necessarily the most important one) in the shift from energy-deficit to energy-surplus states of economic activities which are directly or indirectly vulnerable to rising energy prices.[6]

Population movement in this country had been westward and southward long before 1970, and the trend would probably have continued even if the real price of energy had remained unchanged. Some reasons that are psychological

HIGH ENERGY COSTS: UNEVEN, UNFAIR, UNAVOIDABLE?

rather than economic defy quantification: relatively undeveloped areas have something of a "frontier" appeal, and sunny climes are another attraction. Other factors that have little or nothing to do with energy are the presence of nonfuel natural resources, access to markets (and the growth potential of such markets), the availability of appropriate labor—either general workers at competitive wage-scales, or specialized employees whose reasons for clustering in an area may be independent of the enterprise—and others. The second and third considerations are more or less people-oriented, however; so later in this section we will need to look at energy-related but nonprofessional motives for interregional shifts of individuals.

Rounding out the discussion of industrial movements, one might speculate that one reason these calculations failed to uncover a larger and less equivocal reaction to generally rising energy prices is the caveat which also applies to analyzing effects on the poor: all energy is not the same. The various fuel forms are not all readily or completely interchangeable. In addition to that, government price regulation itself has not been evenhanded. As a result, the effect of demand for West Virginia coal may not be truly analogous to that for other U.S. energy resources. British thermal units do not convert directly into profits.

Another factor is that state borders are not the same as national frontiers. State-imposed "severance taxes" on mineral resources must be paid by local users as well as outlanders; otherwise they would be internal tariffs, forbidden by the U.S. Constitution. Yet this means that the cost advantage to be gained through the shift of an enterprise from an energy-poor to an appropriately energy-rich state may still be limited to the difference in transportation cost (which assumes less relative importance as the price at the extraction point goes up). Finally, the period contemplated in the analysis may be too brief to reveal gross shifts. Different factors are associated with different lengths of the adjustment process. Demographic changes may be generational, the life of capital goods implies a lead time of twenty years or more, availability of labor skills may have only a short lead time, and so on. Changes in energy prices will interact with all of them. Thus, interpretation of gross data is doubly difficult.

Federal regulation has also played a role. For most of the 1970s, U.S. price control had a specific perverse effect on industry within states that produced natural gas. Because the price of this fuel could not be controlled when it was sold to intrastate customers, it rose far more rapidly for users near the production sites than it did in other parts of the country.[7] The other side of that coin is that shortages developed *outside* the producing states; and to a limited extent this favored the location of some industries (which needed uninterruptible supplies of natural gas at almost any price) *within* the boundaries of states with gas fields.[8] Still this was not much of an incentive to interregional change—because such industrial location had already largely taken place.

Availability and price have long been important factors in location

Both availability and price of energy have, of course, been important factors in commercial and industrial location for a much longer time than is broadly recognized. Those who see the relationship as a new one have simply taken too narrow a definition of "energy." In the days of sail, for instance, the situation of settlements along lakes or seacoasts could be ascribed directly to what were then the fastest, cheapest, and most comfortable sources of transport energy—wind and water. The earliest U.S. mill towns tapped another source of energy—falling water. The transport costs of fuel are, of course, an important locational factor which, in turn, affects the profitability of industry in specific locations. On the whole, steep price increases in energy will tend to downgrade their impact.

More recently, regional variations in the price of energy simply did not attract the attention of many economists; but this does not mean they were insignificant. For example, an unpublished analysis by the Federal Energy Administration tested the possible effects of narrowing regional price differentials—with astonishing results.[9] Even during the 1963–72 period (when energy was thought to be a relatively modest element in the equation), a move to even out energy prices on a nationwide basis apparently would have increased or decreased the growth rates of some states by 60 percent.

At first, the FEA analysts would not trust their own results. As conference participant Donald M. Smith commented, "[We] began throwing in other variables, hoping that they would click in the correlations with growth rates and destroy the relationship with energy prices, and we used things that had been brought up—like government burden, tax breaks, and that kind of stuff, or unionization, inflexibility of wages, and some institutional things, and could not destroy that relationship."

46

It appears that Texas-style growth has owed even more than was thought to the energy factor—not to the simple fact that oil and gas were there, but because they were cheap relative to the North. It was so even when energy was thought to be cheap generally.

Where does all this leave us in terms of justice to various regions of the United States? According to the regional per-capita income trends, there has been convergence.[10] Except for the West Coast (where per-capita income in each of the two SFRs is above the national average and pulling farther ahead) regional deviations from the nationwide mean have narrowed since 1969. The traditionally wealthier northeast quadrant of the country has slipped, while the Midwest and the South have climbed toward parity. Uneven national endowments with modern energy resources and the recent turbulence in their costs may have contributed to this leavening effect; but the evidence is not convincing and the implications for national policy are uncertain.

Regional per-capita incomes are converging

Energy prices and household location

Turning attention from changes visible on the production and employment side of the ledger, we must ask next whether the role energy plays in a typical family's cost of living may have affected population and income shifts.

Earlier,[11] we mentioned the striking differences among the major census regions in the prices householders must pay for various sources of residential energy. Because, as has been shown in chapter 3, space heating is typically the largest consumer of energy in the home, we expect Northerners as a class to have the greatest residential energy needs. We also remember that natural gas is still the most economical home-heating fuel, but not available in equal measure to all regions. Table 4-1 shows the combined effects of these circumstances on residential energy bills, drawn by Beebout and coauthors from the 1978–79 NIECS data (see table 3-1).[12]

In analyzing the data, one must keep in mind that they reflect few of the changes caused by the Natural Gas Policy Act and that the effects of this legislation will substantially alter price patterns.

If residential energy cost were the sole criterion for location (which it is not), table 4-1 would suggest a strong incentive to follow the crowd in transferring residence southward and westward. Residential energy economics seems to offer a bonus of from 1 to 4 percent of annual income in return for doing so. Or—to look at it from another viewpoint—the higher cost of heating oil has been a noticeably heavier burden on the Northeast as a region. Therefore, some have argued, the Northeast deserves some special consideration from the rest of the nation.[13]

It cannot be overemphasized that cold climate offers only a partial explanation for the differential impact. The pattern is complex. It involves both differences in energy forms and in prices. Because of air conditioning, the South uses a lot of electricity; yet its relatively low rates and its lower requirement for space heating both help to keep its overall residential energy costs below the national average. When all sources are combined, Southerners use the smallest amount of energy in their homes—only a bit more than half as much per residence (in British thermal units) as those who live in the North Central region. In fact, no other region tops the North Central, where average household consumption is 180 million Btu per year. But the Northeast pays *more* for a little *less* than that; as table 3-1 showed, it tops the nation in unit price for all four key sources of energy. Finally, the bulk of Westerners are doubly blessed. Their

Table 4–1. Regional Differences in Average Annual Residential Energy Expenditures Per Household, for All Energy and by Source (April 1978 to March 1979)

Region	Expenditures on all fuels		Expenditure (in dollars) by users of			
	Absolute (dollars)	As a percentage of household income	Electricity	Natural gas	Fuel oil and kerosene	Liquefied petroleum gas
Nationwide	724	8	390	314	501	241
Northeast	887	9	345	335	594	175
North Central	821	8	394	402	559	362
South	674	7	480	250	291	211
West	469	5	283	216	382	278

Source: Extracted from the conference paper by Harold Beebout, Gerald Peabody, and Pat Doyle, entitled "The Distribution of Household Energy Expenditures and the Impact of High Prices," based on statistical data from the 1978 National Interim Energy Consumption Survey (NIECS), table 7, p. 18.

HIGH ENERGY COSTS: UNEVEN, UNFAIR, UNAVOIDABLE?

climate is relatively mild, and the prices they pay for their two major energy sources (electricity and natural gas) are the lowest for any region.

Until the NIECS data on gasoline consumption become available and are analyzed, we can only speculate about the degree to which driving costs now shade this picture. Travel distances tend to be larger in the South and West; according to a recent calculation for 1975, gasoline consumption per household varied from a high of 2,222 gallons in Wyoming to a low of 864 in New York State.[14] But if the 1980 FOMAC estimates hold true (see table 3-3 and related discussion) average transport energy expenditures vary sharply also with other factors (such as income). And, intuitively, we might suspect that home-to-work commuting (a reasonable parallel to the energy necessity of home heating) depends at least as much on the choice between urban and suburban residence as it does on regional location. That would tend to mask any relative advantages or disadvantages for different locations. Cities and suburbs are everywhere.

For that matter, so are rural areas—whose population growth during the 1970s was about three times as rapid as cities and close to the percentage growth of suburbs. According to one observer, the flow of people and jobs into rural communities is being governed by a unique set of factors. Brady J. Deaton of the Virginia Polytechnic Institute reported to the conference on his research in the Appalachian Region. It suggests that the psychic rewards of a more natural and relaxed quality of life in rural areas are considered important enough by those who live there to offset a 30 to 40 percent differential in monetary wages, even when partially offset by lower living costs, from comparable jobs in or near a large city.[15] If this is true generally, one might expect an accompanying migration of certain types of industry to take advantage of the growing pool of low-cost labor.

Deaton believes that these new rural enterprises tend to be less sensitive than most to energy costs. Nonmetropolitan area employment in manufacturing outstripped the sluggish nationwide pace of the 1970s by registering a significant 17 percent increase. But the smaller absolute number of service-performing jobs rose in rural areas at an even faster rate—42 percent. And Deaton's sampling of one state (Tennessee) characterizes the rural manufacturing firms as those for which wages and salaries are much more important than either energy or transportation costs. His impressions support the

Growth in nonmetropolitan employment outstripped the nationwide pace

Appalachian Regional Commission's projections for continuing population growth, unhampered by developments in energy pricing; and the only warning flag he raises is in connection with the loss of markets for manufactured products (such as textiles and clothing) to competitors overseas, where labor costs are even lower.

For all its fascination, however, the energy outlook for rural communities remains a sidelight in the national picture. We are still essentially a metropolitan population—either city-dwellers or suburbanities—and the question at hand is whether changes in energy price cause equal burdens for both.

The urban–suburban energy tradeoff

Indeed, as energy price began to rise and the scarcity of one or the other energy source became an event to be reckoned with, one of the hotly debated issues was the possibility that suburbanites would be penalized vis-à-vis inner-city dwellers, and that the outward migration of city people would be not only halted but reversed. The conferees considered a simple model to gauge the differential impact of rising energy costs on those who live at different distances from a "downtown" employment site, based on the hypothesis of the real cost of gasoline rising by $1.00 per gallon above the 1977 level and heating oil rising by 75 cents.[16]

The model contrasts a high-energy-cost residential situation (a single-family detached house in an auto-oriented suburb, where four out of five family members drive, and the one-way distance to work for the one or more employed members is 15 miles) with a low-energy-cost example (a multifamily structure served by public transit and only five miles from the principal workplace, with two drivers out of five). The cost difference between these two situations in direct energy outlays resulting from the assumed price rise would be nearly $900 per year. This would be divided almost equally among (1) gasoline for commuting to work, (2) gasoline for other auto travel, and (3) heating–cooling costs.

Naturally, these two examples are extreme cases. Census data suggest that a typical suburbanite household would pay an energy premium of slightly more than $250 as compared with its in-town counterpart. If one were to capitalize this item of continuing expense, the relative asset value of urban residential property would rise by virtue of its being more economical to occupy.

This opens an intriguing perspective: a net shift of these dimensions in property value could be huge in the aggregate if it actually took place. We might be talking about a relative gain of more than $100 billion in asset value for U.S. cities over suburbs; and—through happenstance—the principal beneficiaries *might* be the relatively poorer groups who are now clustered in the suddenly desirable city core. However, because they do not, by and large, own property and because many intermediate adjustments are likely, no such bonanza for the lower economic classes should be expected. Nevertheless, effects of this magnitude are too large for policymakers to ignore. They have potential ramifications for urban planning as well as for poverty-relief programs; and while the adjustments hypothesized in the following paragraph are valid for housing anywhere, the greater availability of energy-efficient multiunit dwellings gives the matter special importance in cities.

During the first year or two, those who occupy relatively low-energy-cost housing will have advantages which they are unlikely to perceive; yet landlords for high-energy-cost residences will suffer in competition with those that are more desirable in terms of energy-cost. Gradually, both rents and sale prices for energy-cost-favorable locations will rise to compensate, as tenants and buyers adjust their demands. Those who cannot afford higher living costs will settle for less space or less convenience at higher unit rates—and the adjustment will be more difficult if they cannot find some way of applying energy more productively (for example, through more efficient home-heating arrangements or personal autos that carry them more miles per gallon).

In the longer run, inner-city areas may face a piecemeal process which has come to be called "gentrification." Many close-in neighborhoods are too run-down to provide substitute housing for erstwhile suburbanites until capital improvements are made to individual units. Ironically, people who lack the cash to realize their *relative* energy advantage feel only the pinch of *absolute* price rises; and potentially valuable property is even abandoned outright at times. As parts of these neighborhoods improve, however, tax assessments and rentals are likely to rise throughout them. Poor residents, depending on whether they are owners or renters, may be squeezed out slowly—with or without reaping the full

A relative gain of more than $100 billion in asset value for U.S. cities over suburbs

economic benefits to which their convenient, close-together houses might theoretically "entitle" them in a period of rising energy costs.

The model's author makes no pretense that this attempt to develop an urban versus suburban model for energy-cost impacts has produced a very sophisticated one. Exceptions are plentiful. People who live in metropolitan outskirts may not work within the city limits at all; they may commute along beltways between adjacent suburban locations. Some of the most recent housing developments in the far fringes of cities may also be energy-efficient highrises or group homes. Rising energy prices change people's attitudes toward ride sharing and wearing sweaters. Things change too: cities themselves spread out (or conceivably contract). New housing is built. New work sites appear.[17] And, as Ellis Cose of the *Detroit Free Press* remarked, "Differences among some neighborhoods are so striking that many, even if they are close in, would not necessarily have their value bid up in the short run because gasoline became more expensive."

Future regional and urban trends

Still, it seems safe to venture some general observations about future trends that economic rationality might dictate. Flights to the suburbs should slow down, if not reverse. Whether cities are in regions that are now booming or declining, those that hope to prosper would do well to stress high density, good mass transit,[18] lots of multiunit dwellings, and a high proportion of energy-efficient homes (regardless of age). In areas where district heating appears practical, it offers added advantages.

It is doubtful that any single economic model could analyze and project regional differences in energy-price impact at the same time it integrated urban–suburban variations. The difficulty of such an assignment should not discourage efforts along those lines; but meanwhile we can risk some regional projections on our own, keeping in mind two harsh energy facts:

1. *The natural gas price honeymoon is already scheduled to end.* Because world oil prices have risen much more rapidly than Congress anticipated when it passed the Natural Gas Policy Act in 1978, even the gradual escalation of wellhead prices contemplated in the legislation for newly discovered

sources of this fuel is almost certain to leave them far short of parity with oil when the time comes for substantially full decontrol in 1985. Barring greater interim adjustments, at that point the delivered price of natural gas for intrastate use would take a large jump—some believe at least double— overnight. Outside the producing state, the average delivered price will rise less, but still substantially.[19] Long-term supply contracts and "rolled-in" pricing will cushion the effects more for some consumers than for others.

2. *Residential unit rates for electricity (which have risen only moderately in real terms) could rise precipitously in some service areas by the late 1980s.* Utility orders for new central station generating plants have been eliminated or postponed for a great many reasons, not the least of which is the poor capital situation of the utility industry. Even though demand is growing much more slowly than it did a decade ago, a greater than anticipated switch to electricity—for example, from newly expensive gas—or con- tinued delays in putting new generating capacity on line could make reserve margins in some areas of the country uncomfortably slim within a decade—even in what have been off-peak periods.[20] Moreover, a hodgepodge of circumstances may bring about the premature retirement of some generating plants (rather than let them be converted uneconomic- ally to coal[21]) at the same time other old coal-fired plants are being kept in inefficient service beyond a reasonable age (because of new-plant delays). The result could be to place more reliance on banks of small combustion turbines, which involve relatively little capital and short lead time for orders but which have high operating costs. Fuel cost adjustment clauses in local electric-rate structures will assure that these increases are felt promptly by customers. Even if baseload plants are added eventually, interim inflation and the high cost of capital funding will cause sudden swells in the rate base—equally bad news for consumers in the longer run.

Some specific speculation surfaced at the conference. A state like Minnesota appears to be headed for trouble. It has cold winters, and its homes use a lot of natural gas (destined to rise steeply in price) for heating. Florida faces problems too, but for different reasons—its heavy use of electricity for air conditioning and the fact that its burgeoning population could strain generating capacity. The Northwest, which has long been favored by cheap hydroelectricity, seems to have no place to turn for large additional blocks of power except steam-generating plants—which means very sharp increases in electricity prices there as well.

The southwestern United States is also on the new endangered list, for several reasons. Industry and commerce there have been built to a considerable extent on cheap natural gas—a fuel whose U.S. market price is destined to rise rapidly, and sooner or later to approach some grades of oil. The shock to the Southwest is bound to be even sharper in comparative terms because of a shift in the components of gas price to the user. The big regional differential has been transport costs (which constituted a large fraction of the high price Northerners were paying for this fuel). Now, suddenly, the *wellhead* price assumes a greater importance. Not only will Arizonans be paying more for gas in absolute terms in the future; their relative cost advantage over New England in using this fuel will be a much smaller percentage than before. And, because of population growth, southwestern states will also be compelled to add new electric-generating capacity at a faster rate than most of the country, forcing electricity costs upward as well.

In the course of the conference, Mary Procter, of the Office of Technology Assessment, advised the participants to accept the results of sophisticated statistical analysis with caution. Is it not intriguing, she commented, that a state like California, which statistically appears relatively well-off, has an abundantly staffed state energy bureaucracy, while a state like Massachusetts, which appears all but well-off in matters of energy, makes do with a bare handful of state energy officials? Perceptions may not always match the facts.

If we look for ways to share the burdens of higher energy costs fairly among regions, we have the dilemma of looking back over the 1970s or ahead to the rest of the eighties. Should we try to redress past grievances or forestall imminent ones? Or—perhaps—should we admit that there is no need to tie regional equity considerations to national energy policy?

As in the case of the poor, this approach is not an excuse to shirk federal responsibilities. Some states may need outside help. A new system of revenue sharing is one possible avenue that will be explored briefly in the next chapter.

Yet separating the two issues of "energy for all the states" and "justice for the respective individual states" is most conducive to finding acceptable methods of dealing with both.

Finally, it must be said that the focus in state-related energy issues, at least as far as the public is concerned, has been less on those discussed above than on physical impact and associated costs and benefits. There are important and legitimate arguments over the extent to which specific areas should be opened to exploration, especially for offshore resources, over water depletion to permit synfuels development or coal transport in slurry pipelines, over location of nuclear waste repositories, over sharply rising transit of coal trains. These and similar regional aspects of present and future energy-connected problems have dominated the debate, but they were touched on only lightly during the conference because it focused on the direct consequences of rising energy prices.

Notes

1. *A National Agenda for the Eighties,* Report of the President's Commission for a National Agenda for the Eighties (Washington, D.C., GPO, 1980) pp. 64–72 and 165–168. For a formal dissent incorporated within the published document, see p. 139.

2. Quoted and discussed by Richard Corrigan and Rochelle L. Stanfield in "Rising Energy Prices—What's Good for Some States Is Bad for Others," *National Journal* (March 22, 1980) pp. 468–474. While data are flawed in that electricity generated by sources other than fossil fuels is included in consumption but not in production, the appropriate correction would not affect the broad grouping, except perhaps for one or two states with high hydro generation.

3. A question that deserved attention but was not considered is the extent to which income from energy production remains in the state in which production takes place.

4. No statistical significance test was applied to these figures. But the fact that the two groups moved uniformly up or down lends credibility to the interpretation.

5. One of these consists of Louisiana (406 percent), Texas (213 percent), New Mexico (338 percent), Oklahoma (247 percent), and Arkansas (43 percent). The other includes Utah (100 percent), Colorado (77 percent), Wyoming (405 percent), Montana (229 percent), North Dakota (119 percent), and South Dakota (39 percent).

6. "Shift" need not imply physical relocation. A net switch in employment may result from a decline in employment for one region due to attrition and a simultaneous expansion in the labor force of another through independent growth.

7. For instance, Miernyk's state-by-state price data (see table A-3, p. 95) shows that in 1970 the price of natural gas for all three categories of Texas users—residential, industrial, and commercial—was below the U.S. average. By 1978, all three Texas rates had risen above the national average.

8. An obvious example lies in certain petrochemicals.

9. Bill Huntington and Don Smith, "Energy Prices, Factor Reallocation, and Regional Growth," Discussion Paper EIA-76-63 (Washington, D.C. Regional Impact Division, Office of Macroeconomic Impact Analysis, Federal Energy Administration, Oct. 26, 1976).

10. William H. Miernyk, "The Differential Effects of Rising Energy Prices on Regional Income and Employment," paper prepared for the conference.

11. See discussion on pp. 21–23 of chapter 3; and especially note the regional breakdown of various energy prices in table 3-1.

12. U.S. Department of Energy, Energy Information Administration, *Residential Energy Consumption and Expenditures, April 1978 Through March 1979*, DOE/EIA-02075 (Washington, D.C., DOE/EIA, July 1980).

13. More recently, and spurred especially by the heat-wave-triggered deaths of the torrid 1980 summer, an analogous argument for assistance in space cooling for the Southwest has been advanced.

14. David L. Greene, Oak Ridge National Laboratories, cited by Reid T. Reynolds in "The Demographics of Energy," *American Demographics* (June 1980). It is somewhat surprising that the Pacific Coast region ranks with New England and the upper Midwest at the low end of the scale for gasoline consumption per household. The heaviest use is in Texas, Louisiana, and all other states west of Kansas.

15. According to a paper prepared for this conference by Paul W. Barkley, of Washington State University, they have also paid a long-standing penalty in higher energy costs. Barkley said that as far back as 1960 a typical rural household was spending 10 percent of its income on all forms of direct energy while the national average was about 6 percent. This included larger amounts of motor fuels and often a disproportionate share of bottled gas, which costs more than either natural gas or fuel oil for heating.

16. The model was set up by conference participant Kenneth A. Small.

17. It is conceivable that the Reagan administration's promised incentive for "free enterprise zones" in central cities could provide stable focal points for current residents who wish to revitalize their own neighborhoods. On the other hand, they might merely speed up "gentrification."

18. The energy conservation potential of mass transit is widely misunderstood. Marc H. Ross's calculations in chapter 5 of *Energy in America's Future: The Choices Before Us* (Baltimore, Md., Johns Hopkins University Press for Resources for the Future, 1979) indicate that a half-full bus uses only slightly less fuel per passenger mile than a 20- to 25-mile-per-gallon auto with a driver and one rider. The capital cost of some fixed-rail systems makes them economically unjustifiable unless social values are weighed heavily. Mass transit is an important factor, however, in encouraging high-density, energy-efficient living and working space while avoiding intolerable traffic congestion and a need for enormous parking facilities.

19. The estimate presented in *Reducing U.S. Oil Vulnerability: Energy Policy for the 1980s* (Report to the Secretary of Energy prepared by the Assistant Secretary for Policy and Evaluation. November 10, 1980, p. II-B-9) is for a rise of one-third.

20. Moreover, "off-peak" periods themselves may become less common. Already, peaks are being shaved, and the traditional distinction between "summer-peaking" and "winter-peaking" utilities is blurring in many cases though, on a national and regional basis, the industry's trade organization, the Edison Electric Institute, foresees no significant changes in the relation between summer and winter peaks through the end of the decade. This may not hold true on a local basis. As an example, the Tennesee Valley Authority has long had its highest demand during winter months because its low rates had encouraged electrical heating, but it now appears that this will be matched during the 1980s by summer peaks caused by greater use of air conditioning. In the Baltimore–Washington area, on the other hand, the difference in size between the customary summer peaks (based on air conditioning) and new cold-weather peaks (based on the rapidly expanding use of heat pumps) is narrowing. It too may vanish during this decade. Existing capacity is used more effectively, but downtime for maintenance becomes more difficult to schedule when demand is more evenly distributed over time or when two distinct peak periods must be anticipated each year.

21. See *Reducing U.S. Oil Vulnerability: Energy Policy for the 1980s*, pp. IVF-7 *ff*.

5

Ways to ease the burden

Despite difficulties in quantification, it is obvious that rapid rises in energy prices hit different people with widely differing severity. There are variations by income, occupation, location (in both the macro and micro sense), fuel habit, and so forth. A question that remains is whether there are ways of sharing the costs of our current energy problems more fairly.

This question may be asked without suggesting that the *status quo ante* (say, prior to 1973) represented perfection in distributive justice or that our target now must be to distribute the cost, in some sense, equally. The foregoing pages should have demonstrated that while the poor, however defined, have been penalized disproportionately by rising energy prices, the sources of inequality in effect are so numerous and so entangled with one another that no computer program in the world could ever develop a formula to equalize the impacts on a personal level. The proper aim is more limited. We suggest that it is to ease two types of burdens: those caused by (1) higher prices, and (2) the uncertainty that necessary energy will be physically available when demanded. To be somewhat more specific, we might define the goal as making burdens at least tolerable to those individuals, enterprises, and areas which otherwise might find high prices or energy inadequacies (or both) threatening their economic survival. Even this is a difficult undertaking.

Three basic approaches

Dozens of different schemes have been proposed, but they can be categorized among three basic approaches:

1. To try to sit on all domestic energy prices and to mandate allocations of all energy forms
2. To set up differentiating systems of price and supply for designated groups of energy consumers
3. To compensate—in cash or kind—those who (according to various criteria) suffer most or who appear least able to bear up under the stress.

The first approach—price control with allocation—has few remaining advocates, and disappointment with policies of the seventies that incorporated some price-fixing and allocation elements has further diminished their number. Here, in brief, are the arguments against such an approach.

Price management produces its own inequities and inefficiencies

If U.S. prices are held below the world price, we will be consuming energy in uses where the value it contributes to national well-being is less than the value of goods and services we must ship abroad to buy the extra oil consumed. Thus we are subsidizing imports. Similarly, because the production incentive is lower, we will produce less energy of all kinds here, even though, again, the goods and services we would have used up in producing that energy domestically would be less than the amount we would have to give up to pay for imports. In these and other ways—such as stimulating demand by an artificially low price—the false signals sent by controlled prices lead to economic waste and a reduction in the potential level of the nation's well-being.

Bureaucratic unwieldiness apart, to bring about the massive control network involved in allocating energy resources without price mechanisms would be tantamount to an institutional (if not political) revolution. This is less likely than ever since the 1980 elections. Short of wartime conditions, it is inconceivable that the nation would voluntarily accept the authoritarian governmental measures which would be needed to enforce low energy prices and a complete allocation system.

Prices have risen unevenly for different energy sources and for different parts of the country. Thus, rolling prices back would not have a uniform impact either.

Some consumers would benefit far more than others: and what seems most unfair is that the smallest rewards would go to some of those who had made a serious effort to adjust to the earlier price rises (for example, by equipping homes and businesses with capital-intensive alternative energy systems, many of which make no economic sense except as a means to save high-cost fuel).

The second approach to easing the burdens of energy price increases and scarcities—differentiating among groups of consumers—makes no pretense of treating all individuals or areas or energy forms evenly; in fact, it moves intentionally in another direction. This approach aims at helping not "the consumer," but only those consumers who (by some agreed standard) need help. It does so by setting up special ground rules for the way in which they acquire energy.

"Lifeline" utility rates are an example. The theory behind them is that there is some minimum amount of either electricity or natural gas which any consuming household requires for subsistence-level survival,[1] and that the charge for this amount should be well below the average unit rate. This clearly reverses the traditional custom of using "declining block rates"—according to which rates are supposed to relate to the actual costs of providing service and therefore to decline with rising consumption, owing to claimed economies of scale.[2]

It is instructive in this context to look at the utility that for much of its existence has been considered a "model" for responding to social objectives. The Tennessee Valley Authority has evolved what amount to lifeline rates. It has done so by an ingenious, if somewhat tortured, interpretation of the part of the TVA Act which provided that hydroelectric projects should benefit "the people of the section as a whole and particularly the domestic and rural consumers...." A paper contributed to the conference by Robert F. Hemphill, Jr., and Ronald L. Owens reported that the system's relatively low-cost hydropower is considered as being used exclusively to supply the basic needs of residential customers, while commercial and industrial rates are based entirely on the higher costs of nuclear and fossil fuel-based generation. Furthermore, it so happens that hydro capacity is roughly equivalent to supplying 500 kilowatt-hours per month for each residential consumer in the valley, deemed to represent basic needs; thus householders in the fall of 1980 paid only 3.11

cents per kilowatt-hour for usage up to that amount, but 3.55 cents per kilowatt-hour for all consumption beyond it.

Lifeline rates undoubtedly have popular appeal

Lifeline rates undoubtedly have popular appeal, but the arguments supporting them do not stand up well under closer scrutiny, largely because differentiating by categories is a very gross approach. It is true that such rates provide a relative advantage to small users, because any dollar amount trimmed from the "first block" represents a larger percentage of small than of large total bills; but in fact a reduction in this price-block is shared by all residential consumers to some extent.[3] Furthermore, small utility bills do not necessarily indicate poor customers—just as big bills are not confined to the well-to-do. People who are affluent enough to maintain weekend retreats could find their intermittent use of more than one residence subsidized incongruously by a measure that was intended to aid the impoverished. On the other hand, some of the poorest Americans would be helped less than intended, because they tend to use less-efficient appliances in homes that are not as well adapted for energy conservation. In regard to regional inequities, lifeline rates universally applied would be irrelevant; spotty adoption could aggravate them. Nor would they have any effect on the costs of liquid heating fuels and gasoline, which probably have caused the most nationwide distress to date, but which are not distributed by publicly regulated utilities.

There is a school of thought that answers the last criticism by going beyond the conventional definition of lifeline rates to propose that a similar principle be applied to *all* energy sources for the poor. Gar Alperovitz has for some time argued that food, housing, and medical care should all be provided (in addition to energy) in such a way that any American could have access to them—at least to the extent that these commodities and services can legitimately be described as necessities of life. His technique for accomplishing this is unclear but presumably rests on some sort of price control; indeed, his writings on the subject favor substantial stabilization of what he calls necessity prices. He admits that "upper classes might benefit unfairly" from the same price modifications, regardless of how they were applied; but in that case "the solution is obvious: taxes to take back any benefits to high-income groups caused by holding down prices."[4] Continuing inflation, and not just rising energy prices, could easily provide momentum to some variety of such an approach. For the moment, this represents a direction of thought rather than a well-elaborated systematic policy tool.

Aside from pricing policies aimed at preferential treatment of the poor, it is possible to give them special consideration also in terms of guaranteed supply, at least in cases where the supplier is a regulated utility. In many parts of the United States, there are absolute prohibitions against cutting off electric or gas service for nonpayment during especially cold weather. Some southern localities have also adopted similar provisions for electricity during heat waves, when air-conditioning units (or at least electric fans) are deemed essential to public health. Although rules of this type may seem only to postpone the inevitable, they have probably eliminated some cases of extreme, life-threatening hardship. At the very least, they give those in the most desperate financial straits time in which to extricate themselves from an energy price-trap—either alone or with help from public and private agents. In contrast, there is no general mechanism for guaranteeing continued supplies of liquid fuel for home heating except by paying the fuel bill. And there is ample anecdotal evidence at least that the small dealers who serve many of the poor are often unwilling or unable to extend customer credit beyond weekly payments.

The third approach to easing energy burdens involves *direct public compensation* to those who are judged to deserve or need it. Not surprisingly, it loomed large at the conference, because it has been the approach in the mainstream of public policy.

The third approach involves direct public compensation

Indeed, it has been at the center of discussion for more traditional assistance programs on food, housing, health, and all other specific aid programs for the poor. With the Reagan administration showing preference for block cash grants and determination of specifics at the local level, a continuing debate seems assured. The matter is rife with controversies:

- Whether assistance should be given in the form of cash (thus enabling individual recipients to make their own choices as to how it would be spent—even giving them the option of cutting back on energy use in favor of more food or some other good) or in kind (so that public funds would be channeled clearly and fully into solving energy-associated problems)
- Whether assistance-in-kind, when favored, might be the supply of energy

resources (via energy stamps or public payment of fuel bills) or help in limiting demand for energy resources (via conservation actions, such as winterizing, financed directly by some governmental entity or made more affordable through tax concessions)
- And, finally, what the criteria might be for identifying recipients.

The assistance record

A survey of governmental energy assistance to the poor which was in place at the start of the 1980–81 winter[5]—that is, just before the new administration took office—revealed two broad types: (1) direct help in meeting the costs of home heating, and (2) provision of the wherewithal for residential weatherization to minimize home heat losses over an extended period. By now, more than half the states finance some form of energy assistance to the poor on their own, but the states' shares of the total effort are minor compared with that of the federal government. For example, New York State provided an Energy Assistance Credit of $35 which may be claimed by anyone over sixty-five whose gross household income is below $14,000.[6] This aid contrasts in size with federal funds for some households amounting to $400 or more to cover annual fuel bills and capital assistance under federal conservation–weatherization programs which may go as high as $1,600 for a single household. Nevertheless, the nationwide program (which is administered to a very large extent by the states) still falls far short of covering the whole problem.

Aid in cash

Exact data are hard to come by, but Cohen and Hollenbeck estimate that during the winter of 1979–80, only about 60 percent of those whom Congress had deemed to be needy actually received assistance with their fuel bills.[7] Through Fiscal Year 1979, only about 2 percent of the residences eligible by virtue of low household income had been weatherized, although the Department of Energy hoped to more than double that percentage before the end of Calendar Year 1980 by speeding up the rate at which a backlog of appropriated funds was being translated into the actual installation of storm windows, weatherstripping, and insulation. And, of course, there is no specific program at all aimed at helping the poor to bear the increased costs of gasoline, most residential energy apart from heating,[8] or the "embodied energy" which has been a factor in the rising cost of so many consumer goods and services (see chapter 3, pages 29–33). Table 5-1 presents estimates of federal funding for the two existing prongs of the energy aid program. The big jump in fuel bill assistance came as a result of two events in 1979: the Carter administration's move in April to decontrol domestic crude oil prices by October 1981, and OPEC's agreement in June on a dramatic crude oil price increase. Over the years the federal programs have also repeatedly changed their names and shifted responsibilities within the bureaucracy[9]; but those variations are less significant than some general trends to improve the programs' effectiveness.

Certain principles seem to have evolved while crisis assistance was becoming a part of energy policy:

Certain principles seem to have evolved

1. Although the FY 81 legislation contains more guidelines for the states than that of the year before, there has been growing acknowledgment that it is easier to pinpoint needs and develop workable distribution mechanisms at the state and local levels than it would be to insist upon a single, rigid, national system. Each state chooses its own program.

2. Government payments to help cover fuel costs (whether made in cash to the ultimate beneficiary or handled in some indirect fashion) have essentially replaced assistance-in-kind such as blankets and clothing.

3. Although eligibility and benefit levels are determined finally by the respective states, the eligibility ceiling has been raised so that programs now *may* reach all those whose income falls below the Bureau of Labor Statistics' Lower Living Standards, which vary by where people live, according to local prices.[10]

4. According to a complicated formula, the amount of assistance made available now bears some relationship to the specific fuel patterns and prices within an area. States are also required now to vary benefit levels according to the income of the recipients, so there is no "all or nothing" point at which those who accept aid would be severely discouraged from trying to increase their own wage incomes because of a large potential drop in government benefits.

Table 5–1. Federal Appropriations for Two Principal Types of Energy-Related Assistance to the Poor, FY 1975–81
(in millions of dollars)

Fiscal year	Federal energy "crisis assistance"	Federal weatherization assistance
1975		16.5
1976	200	27.5
1977	200	137.5
1978	200	130.0
1979	200	200.0
1980	1,600	200.0
1981	1,850	190.0

Note: Eligibility for assistance in the left-hand column now extends to about 45 million people; for funds in the right-hand column, it is about 35 million. Thus, per-household amounts are bound to be small.

Source: Alan L. Cohen and Kevin Hollenbeck, "Energy Assistance Schemes: Review, Evaluation, and Recommendations," a conference paper; and from a communication of the U.S. Department of Energy, Office of Weatherization Assistance.

Aid in kind

The weatherization program has never quite caught on

As for the weatherization program—which was given earlier emphasis than crisis assistance and was funded on a par with the latter as recently as FY 79—it has never quite caught on. In its earlier years it was ordered to stress the use of volunteer labor or workers whose wages were paid under the Comprehensive Education and Training Act (CETA); and its main intent remained vague to the public. Was it principally a poverty assistance effort? A nationally organized drive to save energy resources? Or an enlightened make-work effort to combat unemployment constructively? Through 1979 only about one-fifth of the funds appropriated via the Federal Energy Administration or the Department of Energy had been used, and a searching review was in order.

A handful of states (generally rural ones with cold climates) had managed to improve the thermal efficiency of more than 5 percent of eligible residences.[11]

HIGH ENERGY COSTS: UNEVEN, UNFAIR, UNAVOIDABLE?

Several Indian tribes (who were funded directly) had done the same for about 10 percent of their qualifying homes. If the national response had been comparable, between half a million and a million homes would have been newly prepared to better stand the impact of OPEC's 1979 price jump. But the record across the country was spotty and, in some cases, pathetic. In the District of Columbia, for instance, winterization had been completed on buildings housing only six-tenths of 1 percent of the clearly poor families who were eligible. About $100 million had, in fact, been spent on the program nationwide through FY 79 to weatherize about 250,000 dwellings.

From this context, the Department of Energy passed the word that it had been authorized to reallocate unspent funds among weatherization grantees and that the same could be done among subgrantees who had not "delivered." Changes in program guidelines permitted payments to hire labor or to engage private contractors. Coincidentally or not, the pace of the weatherization program picked up.

Beyond these pervasive problems, both energy crisis assistance and weatherization assistance have had special difficulty in reaching the estimated 50 percent of the poor who are renters rather than homeowners. A large percentage of tenants do not pay their heating bills directly, but bear the cost of rising energy prices through increased assessments by their landlords. As for weatherization, legislators have been loath to provide capital aid to building owners who may not themselves be needy even though they may have poor tenants.

Renters are hard to reach

In each case the solution chosen was the administratively complicated one of reaching agreements with landlords on an individual basis. In return for government aid in paying fuel bills they have had to lower rents or at least pledge to keep them stable.[12] Similar arrangements were adopted for government-financed weatherization improvements. In practice, landlords have been reluctant to make such agreements. There is no incentive for them to do so, and compulsory participation by landlords in fuel aid assistance programs seems unlikely to be adopted widely. Thus, a means of helping poor renters remains one of the unsolved problems of any assistance program that is tied to the burdens of rising energy prices, though community pressures may stand a better chance of reaching individually acceptable local solutions than attempts at developing and enforcing a uniform national pattern.

Not surprisingly, political dealing has played a notable role in the shaping of energy-equity measures. At the federal level, for instance, allotment formulas have been known to drop the logically defensible factor of heating degree-days (a conventional device for comparing winter fuel needs in diverse areas) in favor of heating degree-days *squared*—a bit of arithmetic hocus-pocus designed to produce a pattern of relative benefits which could command the votes required for passage. As for the home conservation program, it has been linked (again for political reasons) with less technically justifiable grants for what is euphemistically termed *appropriate technology* (that is, solar and windmill systems), whose payoffs lie mostly in the more distant future and are biased toward rural users. The cost of some seven hundred such projects funded by the Department of Energy and the Community Services Administration since 1978 has totaled about $9.3 million—not a large sum, but distracting from more conventional, rapid-payoff residential conservation efforts.

Political dealing has played a notable role

Full compensation: Meaning and cost

Despite the continuing use (or misuse) of terms such as *emergency* and *crisis* in regard to energy-price aid efforts, the reality is that if they are needed at all, they will be needed on a continuing basis, rather than momentarily. And, if costly energy resources are to be used more productively in home heating, a weatherization program for the long haul is more to the point. Quick *ad hoc* programs should give way to an appraisal of the size of the problem and the development of criteria by which to judge possible progress toward managing it.

How big might that job be? The conference discussed an original estimate, contained in the cited background paper, aimed at calculating how much annual heating costs for the poor had in fact risen during the five years ending in the 1980–81 heating season. The computation was based on the 1978–79 NIECS data, but factored in interim adjustments for demand responses to higher energy prices, allowing for the unusually cold weather in the 1978–79 winter, and otherwise making the estimate as representative as possible. It did not, however, eliminate the effect of the rise in the general price level.

The findings are roughly as follows:

1. Among families whose income is less than 125 percent of the poverty line, more than one out of four will have spent over $1,000 on residential energy bills during the 1980–81 period. Expenditures for a small but not insignificant percentage (about one in every twenty-five or thirty) are estimated to be at least twice that much.

2. For approximately one-quarter of these families, the annual direct expense for residential energy has risen more than $500 since the 1975–76 heating season. However, those who have the highest current bills are not always the same households whose costs rose the most. Thus, assistance programs based on price increases would result in a payment pattern quite different from programs based on differences in energy cost levels. In fact, the authors found they could not develop any equation made up of readily determinable factors (household size, climatic region, type of residence, and so forth) that could satisfactorily predict future price burdens for segments of a national sample.

3. Extrapolating the estimate to the larger number of families under the more inclusve FY 81 elibility ceiling, and assuming a demand elasticity of .10 (that is, assuming that energy users cut back their consumption 1 percent for every 10 percent increase in price), it turned out that by the end of the five-year period the aggregate of their *annual* home energy bills was $7.7 billion above what it would have been if energy prices had remained constant. If half of the families in this eligible category were to be "held harmless" from price rises of this magnitude during this period by some form of government assistance (or, to take another tack, in order to reduce by 50 percent the impact of the price rises for all of these poor families), the program would have had to distribute about $3.9 billion during the fiscal year nationwide. That is more than twice what Congress actually authorized. Note also that, if 1972–73 is used as the base year, the estimated cost increase for 1980–81 is $10.4 billion rather than $7.7 billion.

These estimates, however, are subject to a number of qualifications. First, the BLS Lower Living Standard is the most inclusive of the poverty indicators. It comprises about one-fifth of the entire population—roughly 45 million people

The size of the bill: First and second thoughts

(versus the 35 million estimated to be below the "125 percent of the Poverty Line" index, and about 25 million below the 100 percent line). Apart from the fact that there is much shifting over time within any measure of "the poor," is the BLS cutoff on the high side for this purpose? Second, the cost is calculated to "hold the poor harmless" for energy price increases in *current* dollars, that is, including, as noted, the large part of the rise attributable to the higher general price level. Third, no consideration is given to the extent to which the income of the poor has been maintained in other ways. On the other hand, the estimate contains some biases in the opposite direction. For instance, the increase calculated is the increase for one year's costs only (1980–81); thus it fails to reflect the annual costs for preceding years, beginning in 1974. Further, the cost estimate does not include gasoline, the price of which has recently risen quite rapidly.

More than half of the price increase was specific to energy

Overall, the $10 billion is no doubt a highly generous upper-bound cost estimate for one year; a lower one might be no more than half of that amount, depending on the extent to which the income of the poor, comprehensively calculated, has kept up with general inflation, and the degree to which energy price increases have outpaced the general cost of living CPI. As for the latter, the CPI for "energy" had, in early 1981, risen to 410 (1967 = 100); that for all items to 264. Related to 1973 as the base, the overall price index had risen by 100 percent, and that for energy by 230 percent; that is, more than half of the energy price increase was specific to energy. So strong a difference would not likely be changed radically by a more up-to-date CPI base. Table 5-2 displays price trends.

Given all these qualifications, the proper size of fully compensatory payments could be determined only with a good deal more investigation. And perhaps the figure itself—when finally discovered—will not be as meaningful as one might consider it at first. To cite one comment by Randall Weiss, a staff member of the Congress' Joint Committee on Taxation: "Real income, and not the isolated analysis of one price, is the appropriate way to measure changes in economic conditions."

In contrast, Mark Cooper of the Consumer Energy Council of America argued on the other side, that is, for considering energy as a specific case, "Thus, in a dynamic world, we have three factors operating to increase the relative burden on the lower income groups. Their income doesn't keep up. They are

Table 5–2. Consumer Price Index, All Items and Energy Components, 1970–80
(1967 = 100)

Year	All items	Energy[a]	Household fuels[b]	Fuel oil, coal, and bottled gas	Piped gas and electricity	Gasoline
1970	116.3	107.0	107.9	110.1	107.3	105.6
1971	121.3	111.2	115.3	117.5	114.7	106.3
1972	125.3	114.3	120.1	118.5	120.5	107.6
1973	133.1	123.5	128.4	136.0	126.4	118.1
1974	147.7	159.7	160.7	214.6	145.8	159.9
1975	161.2	176.6	183.8	235.3	169.6	170.8
1976	170.5	189.3	202.3	250.8	189.0	177.9
1977	181.5	207.3	228.6	283.4	213.4	188.2
1978	195.4	220.4	247.4	298.3	232.6	196.3
1979	217.4	275.9	286.4	403.1	257.8	265.6
1980	246.8	361.1	349.4	556.0	301.8	369.1

Note: These index numbers are based on a consumer market basket that is nearly ten years old, and thus any changes in relative importance of different items in consumer expenditures are not reflected. The problem is discussed in various parts of this report, especially in chapters 3, 5, and 6.

Source: 1970–79: Appendix B, *Economic Report of the President* (Washington, D.C., GPO, January 1981); 1980: Bureau of Economic Analysis, *Survey of Current Business* vol. 61, no. 5 (May 1981); Bureau of Labor Statistics, *Consumer Price Index*, *All Urban Consumers* (Washington, D.C., GPO, 1981).

[a] Household fuels, gasoline, motor oil, and others.
[b] Fuel oil, coal, bottled gas, piped gas, and electricity.

unable to reduce consumption through behavioral responses because they are already at the minimum, and they are unable to reduce consumption through investment in conservation because they lack the capital."

As for home weatherization, it is possible to estimate the upper limits on the total cost nationwide (although individual costs vary a great deal). A recent evaluation by the Department of Energy of the conservation potential from building retrofits to a cross section of residential structures estimated that a "conventional" job (average cost $800) would reduce space-conditioning requirements on each one by an average of 20 percent, and that a 50 percent total saving might be expected by investing an additional $1,200 on each unit (for an average of $2,000) in more sophisticated retrofits.[13] On the same assumption that at least 10 million poor households still stand in need of such attention, we might anticipate an eventual outlay of at least $8 billion and possibly $20 billion to get the whole job done. Unless the existing program were speeded up dramatically, however, the task would take until well into the next century to finish.

Material presented at the conference raised some interesting questions about effectiveness. It challenged claims made early in 1980 by the department's Office of Weatherization Assistance that the average weatherized structure might save more than 100 gallons of heating oil equivalent annually, suggesting that the amount of weatherization actually being performed was less than that in the model, and that in time weatherizing materials would deteriorate. Mention was made also of a study of weatherized homes in Minnesota that showed average energy savings of only 13.4 percent, explained partly by the fact that the occupants of more than one-third of the homes in the survey had raised their thermostat settings after the energy conservation measures had been taken. Apparently, they chose additional comfort (or what they perceived as comfort) to saving money. Thus, from the standpoint of those who favor government-assisted weatherization as a means of conserving fuel, the result of the program was disappointing.

Toward a coordinated assistance program

Given limited assets, poor families are not likely to be persuaded to make large capital investments in anticipation of long-term paybacks. If the nation as a whole opts in favor of lowering energy inputs by reducing residential energy demand (a ripe prospect, considering the sheer volume of energy that goes into home heating), one element would consist of taxpayers at large footing the bill for improvements in lower-income housing. But, to be most efficient, this move should be accompanied by coordination between programs that pay fuel bills and those that offer free weatherization.

There is also the general argument of energy stamps versus cash. In this old debate, one side insists on some guarantee that funds appropriated to relieve the burden of rising energy prices will be spent only on energy. This might be achieved by issuing stamps or vouchers redeemable only for fuel, or by having the government pay fuel vendors directly. The other side disputes such techniques because (1) they do little to encourage energy conservation, because the beneficiary (and energy-user) is free to overlook the real price of that which he uses; (2) the recipients of such aid are denied the privilege of deciding for themselves how to apportion the economic resources available to them; and (3) the marketplace itself may be unable to reflect the true feelings of the general public about how much energy is worth, compared with other goods and services. The conference heard a report that interviews with recipients of governmental energy aid in several states showed that *they* favor direct payments to vendors. Perhaps the security of knowing that bills would be paid was more important to them than broadening the personal choices they could make.

National versus local norms

Those formulating energy assistance schemes have tended increasingly to admit regional differences in burden. Fuel mix, seasonal requirements, and relative energy prices all vary with location; and even a single state is not homogeneous.[14] Nevertheless, it should be easier to approach equitable arrangements within one state than at the federal level, so it may be fortunate that the new administration has announced its general intention of transferring responsibilities in that direction. Now the gnawing question that remains is whether the respective states will meet the challenge of such a task as the reins from Washington loosen.

Can the states meet the challenge of directing energy assistance to the poor?

There is reason for concern. As figure 3-1 shows, states have not responded with equal conscientiousness to an assistance program of longer standing—Aid to Families with Dependent Children. About 55 percent of AFDC funds come from the federal government, the state pays one-third, and the balance is contributed by local governments; but eligibility standards and the level of benefits are left up to the states. In theory, the AFDC grant is supposed to reflect the actual cost of utilities, rent, food, clothing, and other basic expenses, so some variation by state is to be expected. Obviously, however, some states appear to be either totally unrealistic or just plain stingy.

A state's ability to pay the substantial share in AFDC which is required by law ought to be considered, however. Because that ability varies, the possibility of distributing burden through federal revenue sharing or block grants presents itself. If "getting energy prices right" is expected to benefit the entire nation, it *seems* fair that taxpayers from the whole country should participate in a program that tries to alleviate burdens imposed by the policy on individuals or groups who might not otherwise survive them. How might that be accomplished within the administration's obvious resolve to shift more and more decision making to lower echelons of government?

One step in this direction could be an increase in the level of federal revenue sharing, running recently at only about $6 billion annually.[15] Another would be to modify the formula which determines how much goes to each state. At present, the distribution ratios for outright revenue sharing are based primarily on three factors (population, relative income, and tax effort); but there would be several justifications for changing this if revenue sharing were to become a major vehicle for easing the impacts of rising energy prices.

The full picture is not simple. Its complexity can be easily illustrated. Consider the cases of Louisiana and New Mexico, which rank forty-sixth and forty-seventh among the states in personal income per capita. Both are among the nation's leaders in energy production, so that severance taxes and the rents and royalties on oil and gas brought them $1,099 and $1,459 per capita, respectively, during the six-year period between 1974 and 1979.[16] By the late 1970s electricity and gas rates in these two states had risen for all classes of consumer (residential, commercial, and industrial), but they were still below national averages. If our rough projections in chapter 4 (see page 54) are correct, however, both corporations and householders in these states will be worse off than most during the *next* wave of price rises. Where will that leave the poor of Louisiana and New Mexico, who—as explained earlier in chapter 3—may be expected to suffer with special intensity? As figure 3-1 shows, these two states have been among the most niggardly in doling out AFDC benefits.

Personal income averages are relatively poor estimates of burden

As things stand now, Louisiana and New Mexico are favored by the revenue-sharing formula because their per-capita income is low. Yet personal income averages—by themselves—are a relatively poor estimator of the burden from high energy prices. They give no clue that these states had been less hard hit

HIGH ENERGY COSTS: UNEVEN, UNFAIR, UNAVOIDABLE?

than most, nor that they are in for tougher sledding than many in the future. Nothing in the formula takes note of the fact that oil and gas deregulation constitute as much of a "windfall" for high-production states as they do for energy companies. On the contrary, the higher severance taxes which the energy problems of the 1970s have emboldened states to levy on their outputs now bring them even more money as an extra bonus in federal revenue sharing. According to the third factor in the formula, such states are credited for additional effort to tax themselves.

Elements of a comprehensive program

There is more than one way to cut through this confusing thicket. Taking a fresh look at everything that is involved, however, a new form of creative federalism might offer a reasonable path:

- A new revenue-sharing formula could take into account some reasonably up-to-date assessment of living costs within each state (see chapter 6 for a further discussion of this point). Energy suffuses all activities in our society to such an extent that changes in its price would inevitably be mirrored, even if a bit belatedly.

- Severance taxes may be justified as a means of raising sufficient revenues to compensate for environmental damage which results from energy production.[17] Viewed in this way, they represent an internalization of environmental costs (which allows market prices to curb demand fairly) and a minimally just reward to the states for cooperating with a national policy of increased energy supply despite some inconvenience. However, severance taxes on the production of energy resources are paid chiefly by those *outside* the exporting state, so they should not be considered "tax efforts" and should not entitle a state to a larger portion of shared federal revenue.

- Each state government would decide how to help the disadvantaged within its own borders withstand the buffeting of rising energy prices. These governments should be in a better position than any national entity to know what types of programs are most likely to yield a just distribution of benefits in the face of contradictory effects from energy price increases; but spotty performance in the AFDC program suggests that pressure on them should be kept up from at least three directions: (1) citizens of each state ought to be aware that programmatic responsibilities had shifted from Washington; (2) results should be monitored centrally and continuously by a group like the Advisory Commission on Intergovernmental Relations, drawing data from the Energy Information Administration, the Bureau of Labor Statistics, the Department of Health and Human Services, the Bureau of the Census, and others; (3) some residual "clout" to affect state policies should remain with the federal government.

- Although equity policy would be divorced in the main from energy policy, the shared funds to carry it out might still be derived from energy-related taxation. Congress indicated its intention that a substantial fraction of the Windfall Profit Tax was to be devoted in future years to relieving those special burdens on the neediest citizens that might be traced to the higher prices caused by oil price decontrol. Congressional horse-trading might easily insist upon a similar pledge in regard to the natural gas industry in return for full and prompt decontrol there. Revenues from any new federal gasoline tax or duty on crude oil imports in the future could also be shared by whatever formula had been adopted previously.

An energy future for the United States without some continuing adjustment is wishful thinking

The initiative rests with the president and Congress, but an energy future for the United States without some continuing adjustment is wishful thinking. At this point, the problem of energy assistance merges into that of alleviating poverty generally. The merits of each can be argued separately and jointly on general philosophical grounds and on the level of effectiveness and expediency. What further complicates the discussion is that material discussed at the conference was very closely tied to existing welfare schemes and federal–state relations (for example, AFDC, revenue-sharing, and others). The programmatic declaration as well as the budgetary decisions of the Reagan administration strongly suggest that we cannot lightly assume continuity. We are, therefore, leaving the detailed descriptions and analyses of programs that existed in late 1980 to be reflected in the background papers to be published later. Here we merely remark that:

1. There is much to be said in favor of cash grants so as to allow recipients to follow their own evaluation of needs (extra clothing versus extra fuel).

2. In the current climate regarding general assistance to the poor, assistance tied to energy probably has a higher survival chance than if it were merged with other objectives.

3. Given the wide diversity of needs, and the failure so far to devise a formula with satisfactory predictive characteristics, the closer the aid allocation decision can be made to the intended recipient, the less chance for misallocation.

4. Against this must be weighed the added cost imposed on any system that tries to tailor its operations to smaller rather than larger categories and groups; and, unless state performance is closely monitored by the federal government, gross differences in aid amounts, not related to calculated needs, can quickly emerge—as they did, for example, in the case of the AFDC program (see figure 3-1).

5. When the windfall profits legislation was passed, the Congress indicated its intention that as much as half of the proceeds—totaling many tens of billions of dollars during the act's lifetime—should be allocated to assisting those members of the population who were hit especially hard by higher energy prices resulting from decontrol. Intentions do not bind future Congresses, and the topic is bound to arise as the proceeds are used.

No matter how disinclined the new administration might be to fund energy-associated assistance specifically, or approve categorical grants generally, the question of how to reconcile a move toward block grants to states with the specificity of compensation for burdens caused by oil price decontrol will have to be resolved. In such a resolution the political process will be called upon to combine these ingredients:

- *Equity* (tilting the financial burden so as to alleviate the cost toward the lower end of the income scale, taking into account site differences and so forth)
- *Efficiency* (using some mechanism—pricing at market levels apart—to channel the flow of assistance funds at the right time to the right people)
- *Expediency* (establishing a scheme that will be able to garner the needed votes)
- *Energy conservation* (foregoing, at least, moves which gratuitously encourage energy consumption).

To what extent should the wishes of the intended recipients (provided that they can be ascertained with a degree of confidence) be considered? That remains a question on which the conference participants found it impossible to achieve consensus. They did not have to. Politicians do.

Notes

1. Table 3-2 (p. 30) shows that even households with total income of only $5,000 at the time of the NIECS survey were spending around $20 a month on natural gas and the same amount on electricity. The grossest sort of conversion (using nationwide average prices) indicates that at the time of the survey this was the equivalent of about 7,000 cubic feet (roughly 70 "therms") of gas and just under 500 kilowatt-hours of electricity.

2. Regardless of the amount of individual usage, each customer is assessed some amount for the expense of maintaining lines to the house, reading meters, and so forth. This may be listed on the bill separately as a fixed charge or incorporated into the rate for some minimum amount of electricity or gas. Beyond that, however, the unit rates decrease (sometimes in a steplike fashion) on the assumption that economies of scale exist. In practice state-regulated rates are much more complex than this. Moreover, in recent years, rates have begun to differentiate between summer and winter demand conditions and by time-of-day, all reflecting an impulse toward marginal-cost pricing (on the reasonable assumption that demand beyond a certain point imposes different costs, which offset economies of scale, and that added generating units are more expensive to construct). Those intricacies of rate-setting and proposed reform predate the OPEC era and have been leading a life of their own. They will be examined here only episodically.

3. More complicated versions of lifeline rates for electricity in California were described by TVA's Robert Hemphill and by Jan Acton, of the Rand Corporation. Some systems take into account climate, appliance ownership, use of electricity for heating, availability of substitute fuels, and other factors. Hemphill reported, however, that even such fine tuning had not satisfied everybody concerned and that the formulas were due for further restructuring.

4. Gar Alperovitz, "Energy and Inflation: The Broad View," in Ellis Cose, ed., *Energy and Equity: Some Social Concerns* (Washington, D.C., Joint Center for Political Studies, 1979) pp. 3–4.

5. This segment of the report leans heavily on "Energy Assistance Schemes: Review, Evaluation, and Recommendations," a background paper prepared for the conference by Alan L. Cohen and Kevin Hollenbeck.

6. Leonard M. Greene, "Let's End the Traffic Jam of Energy Aid for the Poor," *The New York Times*, July 29, 1980.

7. Cohen and Hollenbeck, "Energy Assistance Schemes."

8. During Fiscal Year 1981, for the first time, states may distribute federal fuel assistance funds to help meet cooling costs in cases where it can be shown that such cooling is medically necessary.

9. Up until 1976, the Office of Economic Opportunity and its successor, the Community Services Administration, concentrated principally on weatherization, although some resources were also devoted to crisis assistance. These efforts were succeeded by the Special Crisis Intervention Program of 1977, the Emergency Energy Assistance Program of 1978, the Crisis Intervention Program of 1979, and so on. Only one-quarter of the FY 80 crisis funds passed through the Community Services Administration, another one-fourth went to recipients of Supplemental Social Security Income via the Department of Health, Education and Welfare, and the other half went to the states in block grants administered by HEW. By FY 81, the Community Services Administration was virtually out of the picture; and almost the entire program was being administered by HEW's successor, the Department of Health and Human Services. The Weatherization Assistance Program, currently administered by the Department of Energy, succeeds earlier programs in the Community Services Administration and the Federal Energy Administration. The authority for weatherization assistance can be traced in turn to the Economic Opportunity Act of 1974 (as amended), the Energy Conservation and Production Act of 1976, and the National Energy Conservation Act of 1978.

10. An earlier criterion for eligibility had been 125 percent of the poverty level. For the distinction between the Lower Living Standard and the poverty index, see chapter, 3, p. 19. For problems with both, see chapter 6 (pp. 83–86).

11. Unlike crisis assistance, the Weatherization Assistance Program is still restricted to households whose incomes are at or under 125 percent of the poverty level, rather than below the Lower Living Standard of the Bureau of Labor Statistics.

12. Payment methods vary from state to state. Besides cash, fuel bill benefits have been made available in the form of vouchers, coupons, direct payments to fuel vendors, or lines of credit established with the suppliers.

13. *Reducing Oil Vulnerability: Energy Policy for the 1980s*, which cites applied research reported by Lawrence Berkeley Laboratories, in *Energy Efficient Buildings: 1980–2000* (draft, July 1980), p. V-F-40.

14. There is the additional complication offered by metropolitan areas which overlap state boundaries. New York, Kansas City, Saint Louis, Philadelphia, Memphis, and Washington, D.C., all draw workers and shoppers from states in which energy and welfare concerns may be at variance.

15. Far more than this amount in federal revenues is directed each year to state and local governments if one counts categorical grants (which carry very specific instructions as to how the funds are to be expended) and block grants (which usually represent a lumping together of the categorical grants in a single field, such as education, so that the states are permitted a bit more flexibility in applying them). Revenue sharing is used now in the strictest sense to apply the federal appropriations which have virtually no strings attached (except for standard clauses guaranteeing against discrimination in hiring). Purists might object to our extension of revenue sharing to cover a program that is assumed to embrace a certain problem area (namely, adjustments to higher energy prices); but the states would retain such great programmatic latitude under this proposal that it seems most akin to the very concept of revenue sharing—regardless of semantic tradition.

16. As calculated by Allen D. Manvel and quoted by George F. Break in "Fiscal Federalism in the 1980s," in *Intergovernmental Perspective* (Washington, D.C., Advisory Commission on Intergovernmental Relations, Summer 1980) p. 14.

17. In the broadest sense, this includes the turmoil and new requirements for service generated by boomtowns, the drain on water resources arising from such activities as shale oil production, and the air–water–land pollution which may result. The role of compensatory payments in persuading localities to accept the presence of energy facilities is treated briefly but thoughtfully in *Energy in America's Future: The Choices Before Us* (Baltimore, Md., Johns Hopkins University Press for Resources for the Future, 1979) pp. 465–468.

6

Where do we go from here?

A highly relevant development since October 1980, when the conference took place, has been the broad debate on the general problem of government assistance to the needy (or, as the new phraseology has it, "the truly needy"). Argument chases argument, and each set of data is quickly countered by another set. As this controversy reached full flower in mid-1981, the specific issue of energy assistance has gradually slipped from clear view. After all, if welfare standards, allocations, and payment mechanisms are to be heavily shifted to the states and local agencies, any drive for national attention and activity in a narrower area of assistance is bound to become less significant. This would be especially true in case block grants become the major vehicle for federal welfare policy or if the nation should adopt a freer attitude toward revenue sharing, as suggested in the preceding chapter.

The most productive view, however, may be to consider policy in evolution and open to movement in various directions. For example, there is the general expression of congressional opinion in the windfall profits tax legislation passed in 1980 that half the revenues collected under the provisions of the act are to be channeled into aiding those especially hurt by higher energy costs caused by price decontrol. While the point has rightly been made that this particular piece of legislation does not bind the present Congress, or future ones, one would nonetheless expect a congressional and public debate on this point. In this debate the questions raised here are likely to be central, whatever the eventual policy decisions.

The need for better concepts and measurements

Though to many they may seem academic (in the sense of being substantially removed from reality), matters of concepts, definition, and measurement loom extraordinarily large in the area of energy burden-sharing. Unless, for example, we have some grasp on who it is that is disproportionately hurt, why, and by how much, compared with some baseline of normality, the debate could bog down on those matters rather than focusing on possible remedies.

Energy assistance programs have concentrated heavily on home heating

To begin with, one of the most obvious difficulties is that energy assistance programs in the United States have concentrated heavily on home heating; and this is understandable. In most areas, some minimum amount of residential heating is essential to health if not life. Furthermore, heating represents the largest single component in the nationwide computed averages for direct expenditures on energy at home. Thus, the costs of heating a house or an apartment are accepted generally as a symbol of energy need (which they are), but also as a proxy for all individual energy expenditures (which they are not).

Indeed, any assistance formula based on "heating needs" would be completely irrelevant to gauging legitimate needs for gasoline, which also ought to be considered. The average U.S. median-income family spends more on gasoline than it does on space heating, water heating, and all household uses of electricity combined. The effects of gasoline price increases on essential auto uses may involve only a minor fraction of the very poor; but this could still include millions of people. And the weight of the burden on the poorest among those who *are* involved might be substantial, but the raw data for estimating these quantities accurately are just becoming available.

Yet even the most systematic, piece-by-piece research on this and all the other components of direct energy expenditure would still leave a huge gap. For most U.S. households, the annual outlay for either heating or gasoline is likely to be less than what it pays out for the hidden energy embodied in all the products and services it purchases during the same period. This indirect component may be the heaviest energy-price burden of all, applying universally to rich and poor across the country. Unfortunately its very existence is generally passed over, possibly because information is still very scanty and the evidence less visible.

Even if, of statistical necessity, we confine ourselves to heating needs, these vary considerably with where one lives—the type of residence as well as the climatic region. Annual weather variations are obviously a factor too. But far greater discrepancies in the absolute costs of heating (as well as in the rate at which costs may rise) depend on the source of that heat.

If governmental assistance in meeting the cost of heating is offered to poor people, it may make very little sense at this time to equalize the amount of aid automatically to two otherwise comparable families who live side by side—if one of these happens to have a gas furnace and the other relies heavily on electric heaters. There are also endless permutations to such generalized examples, based on where one lives. For all but low-volume consumers residential electric rates at one end of the Chesapeake Bay Bridge are nearly 50 percent higher than they are at the other. Along the West Coast, some Californians pay five and one-half times as much for their electricity as Seattle residents do. Different energy sources and different utility companies, viewed in different context by state utility commissions, are among the explanations. If the goal, in the wording of the Reagan administration, is to provide a "safety net for the truly needy," the president and Congress face an almost impossible task in using any national formula to determine who the needy are when it comes to home heating or to estimate how much assistance individual families might need to survive at a reasonable level.

Rates at one end of the bridge are nearly 50 percent higher than at the other

It is for reasons of this kind that a high priority of the federal government ought to be the speedup, reinforcement, and continuing support of the three-year-old effort within the Bureau of Labor Statistics to find out how Americans really spend their money in this new world of expensive energy. This is related to data that are critical to assessing the relative importance of energy prices, and to the concept and measurement of poverty. Nonetheless, policy dilemmas will remain.

Updating the underlying statistics is a high-priority task

Many of the current arguments about the efficacy of the Consumer Price Index are somewhat narrowly focused and may not endure, in that much of the push to end the price-indexing of various benefits comes only from an impulse to limit government expenditures. If a turnaround in housing costs and some other key components led to what looked like a sharp drop in inflation, the upward pressure on the cost of "entitlement programs" in the federal budget would abate and so might the legislative and executive urge to eliminate indexing. Yet our present CPI would still be a flawed tool; and it would deserve to be treated with equal suspicion.

Political figures, advocates for special interest groups, and even a surprising number of professional economists habitually ignore the routine warning by the Bureau of Labor Statistics that the CPI is not a true cost-of-living index. All the CPI does is tell us how much prices have risen on the market basket of goods and services that represented what the average urban American was purchasing in 1972–73.[1] Unhappily, the "lower living standard" of the Bureau of Labor Statistics (mentioned above, because it distinguishes among various regions in reflecting price escalations) is even more out of date. It is based on the 1960–61 market basket. And the Poverty Index is worst of all—being derived fundamentally from consumption patterns of the 1950s.

Does it matter? Indeed it does. For example, in the early 1950s about one-third of average total household expenditures went to pay for food; by the early sixties the share had shrunk to one-fourth and it was barely one-fifth a decade ago. Yet the 1950 pattern of expenditures continues to define the poverty line. Updating is done not by frequent resurveys but by selection of a multiplier judged somehow to reflect reality.

Work is in progress on a new Consumer Price Index

Work is now in progress on a new consumer expenditure survey, to be released in the spring of 1982, and toward a new Consumer Price Index, which will perhaps become available in 1983, but probably not until 1984. Meanwhile, different analysts construct their own special index numbers, and the argument easily shifts from the substantive and judgmental one of measuring impact on the poor to arcane technical ones about appropriate methodology. What is eminently a public task becomes an arena for controversy among economists and statisticans who fill the vacuum.

Similar problems afflict the consumption base that underlies the Lower Living Standard, still reflecting conditions in the early sixties. Then about eight million households in this country relied on coal as their primary heating fuel, while fewer than one million homes were heated by electricity.[2] A market basket of home-energy purchases based on that pattern could hardly be regarded as a reliable key to regional cost differences today.

Changes since the 1972–73 Survey of Consumer Expenditures (which determines the weighting in our current CPI) have probably been less dramatic;

HIGH ENERGY COSTS: UNEVEN, UNFAIR, UNAVOIDABLE?

but the troubling point is that we can only conjecture—again, on a piecemeal basis—what their net effect might be on the actual cost of living so far as energy is concerned. For instance, an examination of the Energy Information Administration's comparative statistics as well as data on private consumer expenditures in the national account statistics suggest the following:

1. Changes in driving habits have lowered the per-capita use of gasoline for personal travel, so that this fuel is assigned a larger share of overall expenses by the 1972–73 survey than it now deserves: there are more vehicles per household today; but the average car now delivers more miles per gallon (according to official data, 14.29 mpg in 1979 as against 13.10 mpg in 1973), and there has also been a noticeable cutback in auto usage since 1979 or so—presumably as a combined result of high gasoline prices and the sluggish pace of the economy.

2. Aggregate residential and commercial use of all petroleum products has also declined in absolute terms since 1973, despite population growth; but sales of electricity to that combined sector have risen by about 20 percent during the same period. Thus, if we try to use the CPI as a cost-of-living index, we exaggerate to some extent the importance of heating oil (whose relative price has risen rapidly), but understate the significance of rising electric rates (which have risen more slowly). Similarly, aggregate residential as well as commercial use of natural gas has remained virtually constant over the past decade or so, despite the fact that a sizable number of home-heating systems have been converted to that fuel in response to the disproportionate rise in oil prices. Apart from some possible improvement in efficiency, this suggests that individual users have become more frugal[3]—a fact of life that one would like to see reflected in the CPI.

Individual users have become more frugal

3. As noted above, calculations of "embodied" energy costs are extremely complex, because they involve the extraction costs of raw materials, manufacturing techniques, and distribution methods as well as consumption habits. They have probably changed (and not necessarily in the same direction) so far as energy intensity is concerned. Furthermore, the mix of the Gross National Product has continued to shift toward

services and away from manufactured goods; and there are literally thousands of popular products today (for example, many electronic items) that did not even exist ten years ago. Once again, the CPI falls far short of reflecting what goes on around us in the sense that, as the basket of goods changes, so does the embodied energy. For lack of better data, past studies of embodied energy have leaned heavily on data from input–output tables, which are often as much as eight years or so out of date.

The inadequate CPI we have now is being used by the Supplemental Security Income and Social Security programs, as well as by a multitude of labor-union wage agreements across the country. Thus, the opportunity of estimating more accurately the differential impacts of rising energy prices is only one of many benefits that should come from the promised improvements, though it would be naive to assume that CPI modifications can have smooth political sailing, given the vested interests in indexing that have grown up around leaving it where it is.

In addition to price index and poverty measurement problems, a third deficiency of the existing data is that they are generally cast in terms of income. Yet, as we have seen, recorded income is not the best guide in measuring need. There is much income that escapes recording, be this from savings or from real estate or other income-yielding assets, or from wealth transfers, in cash or kind. Until and unless we have statistics that more faithfully reflect the full income and expenditure pattern, the debate over whether the poor are in fact keeping up with rising prices will not end.

What kind of aid?

We need to base judgment on a skillful use of available data

Since the best that can be done at present is to encourage preparation of improved data to tell us who is "deserving" of aid, how different households have been affected, and how different regions have fared, we must continue to base some major judgments on a skillful use of the data now at hand. One of these judgments concerns the relationship between general assistance to the poor and assistance based specifically on energy costs. This turns out to be partly a matter of philosophy, partly of tactics. Moreover, the two approaches can, and do, exist side-by-side. That is to say, we do have generalized cash disbursements under various headings while we also employ food stamps, housing subsidies, and other measures.

There are two arguments for wrapping all aid into a single payment: first, it accords best with the nation's general inclination to maximize consumer choice; and second, it is probably easier to establish generally acceptable categories of families and individuals in need of an income supplement than to calculate supplements to specific needs, be they food, or energy, or housing, or whatever. As has been pointed out, these vary enormously and are hard to squeeze into a widely applicable formula or equation.

The case made on the other side (for dealing with energy assistance separately) is essentially a pragmatic one; that is, even legislators and others who are inclined to look at welfare with a jaundiced eye can readily believe the hardships suffered by the poor from rising energy costs and the desirability of relieving them. Thus, there is an argument to be made (and it has become more persuasive since the November 1980 elections) that the chances of accomplishing something in the way of specific energy assistance may be better when pressed independently than when this item is submerged in general welfare. Indeed, the intent expressed earlier by Congress to dip heavily into revenues from the oil windfall profit tax to compensate the poor testifies to the existence of such a preference.

Mechanics of assistance

A second set of decisions would need to be made with regard to the mechanisms by which aid is made available, provided energy-associated assistance were chosen as the way to go. The basic distinction here is between in-cash and in-kind. The latter in turn divides into providing supplies of energy or providing assistance in reducing energy consumption through improved efficiency (for example, insulation). The compelling considerations are (1) that payment in cash leaves the recipient freedom of choice in expending funds and leaves the door open to energy conservation, and, conversely, (2) that payment in kind ignores the recipient's consumption preferences and provides no incentive toward conservation, unless aimed directly at energy-saving or accomplished through in-kind allocations that are tightly set via centrally decreed allocation standards or formulae. While the in-kind aid may be easier administratively—it could be handled through a small number of suppliers—the program still could not escape the difficult job of determining general and specific eligibility. Also, if assistance were to extend beyond residential fuel use—for example, to embrace motor gasoline, and there is much to be said for that—the in-kind help would become difficult to handle.

As for assistance in weatherization, this appears to hold promise only in the long haul; and results so far have not been impressive. There is also a question of social costs and benefits, related to the extent to which housing of the poor is amenable to the relatively inexpensive fixes and, thus, whether dollars spent on improving energy efficiency in housing of low-income groups yield attractive returns. Once broken windows have been replaced and large cracks have been sealed, structural characteristics may present more costly obstacles. Reaching those in rented housing presents a special problem.

On the whole, we believe that if energy aid is separated from general welfare assistance, the cash route has much to recommend it. It is worth repeating that, according to Alan Cohen, aid recipients appear to favor aid in kind, on the simple grounds that aid in cash is likely to be diverted to pay for other pressing needs. If this is so, it may, on reflection, strengthen the case for cash rather than supply of energy.

Back to basic concepts and values

It is obvious that this entire topic is highly value-laden. Popular support of any measures taken will not be rallied merely by reciting statistics. The public *perception* of fairness in policies is as important as the fact that they *are* equitable by some acceptable standard. While public backing can be achieved through skillful exposition of unrelated facts, it would be far better if a consensus could be brought about through fair and straightforward exposition of the issues—by federal, state, and local government leaders; by representatives of business and industry; by teachers and writers; and by members of the various disadvantaged groups themselves. Such a process could begin by trying to clarify how each of us understands the terms involved. Who is "poor?" What does "truly needy" mean? When we talk about "being fair," do we really mean preserving the status quo, or are we seeking some new allocation of wealth and jobs and decisionmaking power and specific material benefits?

We might advance another step if we could reach agreement on a few fundamental concepts:

- In our social–political–economic system, the differences in living standards between the richest and the poorest are less than in some countries which consider themselves egalitarian systems, yet the differences *do* exist and they are great.
- Rising energy prices always tend to hurt most when they hit below the poverty belt.
- There are strong regional differences in energy price (or supply); and these differences can produce genuine regional problems and sharp interregional strife.
- Each distinct energy form has some relationship with other forms, both in price and supply; yet various forms are not readily interchangeable, and often not at all.
- Energy conservation efforts are more helpful in some circumstances than in others,[4] but in general the era of cheap, abundant, reliable energy in this country has ended. We need a new energy ethic, based on this fact alone; and some form of burden-sharing must be part of it.

Looking for perfection will only delay needed action, but good energy policy and good welfare policy can coexist. To see at the core of the energy problem a continuing conflict between equity and efficiency is a severe distortion. Instead, compassion must be informed by cost and efficiency considerations, and the latter must be modified to allow for the selfless impulses of a humane society.

Good energy policy and good welfare policy can coexist

After the RFF–Brookings Conference, Schelling expanded his original remarks for the forthcoming *Proceedings* volume to include four possible scenarios of ultimate consensus regarding energy assistance to the poor. The first of these comes closest to the one we have just offered; but we think that all four of the approaches he envisions are worth capturing here, because any one of them might be defended on both intellectual and emotional grounds:

1. We might accept the principle that the poor should be helped according to need and even develop separate price indexes for the poor

that give greater weight to energy, yet recognize that fuel by itself is a poor vehicle for transferring income. Thus, we might wind up increasing public assistance or making it available to more people as fuel costs rose, but we could still concentrate on incomes, benefits, and the relevant price indexes, rather than on paying energy bills specifically.

2. Another approach might be to try to link the most damaging effects of rising energy prices more closely to the benefits which others enjoy from them—using windfall profits via industrial taxation to reimburse (in cash) whatever group we choose to identify as poor by the estimated amount of *increase* in their prior energy costs. Some inequities would be admitted as unavoidable, but at least people would not be discouraged from trying to conserve energy.

3. Some might take the tack that a government which chooses not to shield its poorest citizens from rising energy prices still has a specific obligation to see that they are not deprived of basic energy needs. Timely payment of their actual energy bills (either directly or through some voucher arrangement) would not encourage conservation; but it might be argued that the minimal needs of the poor are only a small part of the overall energy picture and that if conservation by the poor is the objective, it is better served by government programs to facilitate weatherization or even the modification of home heating equipment.

4. Still another attitude might emerge from a sincere conviction that no energy assistance program targeted only on the poor will ever succeed, but that the stronger constituency of *all* consumers (poor or not) could compel the government to take a more activist role—controlling prices while mandating conservation and other energy policies.

While these four examples spread across a broad segment of the ideological spectrum, there are certainly more extreme positions at either end, not to mention the possible combinations or intermediate positions. Time will reveal what pattern of philosophy and action will prevail. Our strongest recommendation is that the decisions regarding it be made on the basis of the best information available, that the information base be greatly and rapidly enlarged, and that the advocates of different schemes as well as the holders of different philosophies not consider the evolution of a national policy as an adversarial venture. Finally, we should all keep in mind that "the energy problem" is perceived to be less than a decade old. Many adjustments to higher costs and changing availabilities are no doubt in the making, influencing decisions of individuals and institutions even now. Knowing where to interfere and where to merely watch and record is an increasingly important aspect of wisdom in a market society.

Notes

1. The argument here is quite distinct from the now commonplace charge that the CPI errs technically even in this overall task, for example, by assigning improper weight to home mortgage rates.

2. *Residential Energy Uses* (Washington, D.C., U.S. Department of Commerce, 1978), an eight-page pamphlet of maps and graphs surveying changes in fuel consumption patterns over three and a half decades. See especially chart 1.

3. This is borne out by "A Survey of Actual and Projected Conservation in the Gas Utility Industry: 1973-1990" (Arlington, Va., American Gas Association, March 1980) which reported that overall gas consumption per customer in the residential sector declined by an average of 2.7 percent per year between 1973 and 1979. Home space-heating showed an even steeper decrease, with individual customers using an average of 3.7 percent less each year during the period. The American Gas Association ascribed the per-capita reduction to broad adoption of various inexpensive conservation measures (such as caulking and setting back the thermostat) and suggested that the greatest share of gas conservation during the 1980s would probably have to come from more expensive actions (such as installing newer, energy-efficient equipment and insulating).

4. For example, saving a kilowatt-hour of electricity generated by coal-fueled or nuclear power baseload plants during a period of low demand is less significant (both in cost and resources) than saving a kilowatt-hour which would have to be produced by oil-fired turbine peakers.

Appendix A
Some supporting data
from conference papers

The three tables which follow do not have a common theme. Table A-1 supplies the data that underlie figure 3-3, which are basic enough to deserve a separate chapter apart from their tabular presentation here. Table A-2 furnishes the data that underlie the discussion of interstate "terms of trade" in chapter 4. Table A-3 illustrates the wide differences in natural gas and electricity prices, both as between states and according to customer classification. It supports the discussion in chapters 2, 3, and 4.

Table A–1. Household Energy Consumption, Expenditures, and Users by Source

Fuel type	Consumption (trillion Btus)	%	Expenditures ($ million)	%	User households (1,000's)	%
Electricity	2,469	23	29,887	54	76,572	99.9
Natural gas	5,575	53	15,296	28	48,991	64
LPG	327	3	1,663	3	3,124	4
Fuel oil and kerosene	2,192	21	8,623	16	16,919[a]	22
Total	10,563[b]	100	55,469	100	76,608	—

Source: U.S. Department of Energy, *Residential Energy Consumption Survey: Consumption and Expenditures, April 1978 to March 1979* (Washington, D.C., DOE, 1980) as cited in Harold Beebout, Gerald Peabody, and Pat Doyle, "The Distribution of Household Energy Expenditures and the Impact of High Prices," a paper prepared for this conference.

[a] Includes only households in which fuel oil or kerosene is used as the primary heating fuel.

[b] Does not include household use of gasoline for transportation or the use of wood or coal.

Table A–2. Exchange Ratios Between Selected Commodities and Fossil Fuels, 1970 and 1978

Location of producer	Commodity	Quantity	Percentage of U.S. production 1970	Percentage of U.S. production 1978	Value ($s) 1970	Value ($s) 1978	Texas natural gas (10^6 cu. ft.) 1970	Texas natural gas 1978	Texas natural gas %Δ	Oklahoma crude oil (barrels) 1970	Oklahoma crude oil 1978	Oklahoma crude oil %Δ	West Virginia Coal (tons) 1970	West Virginia Coal 1978	West Virginia Coal %Δ
NY	Butter	1,000 lb	4.5	3.5	704	1,141	4.9	1.1	−78	220.7	104.7	−53	88.8	33.6	−62
NY	Milk	5 tons	9.2	8.7	599	1,050	4.2	1.1	−74	187.8	96.3	−49	75.5	30.9	−59
NY	Eggs	1,000 doz	3.3	2.7	393	465	2.7	0.5	−81	123.2	42.7	−65	49.6	13.7	−72
NJ	Processed tomatoes	10 tons	5.6	1.5	416	650	2.9	0.7	−76	130.4	59.6	−54	52.5	19.1	−64
MA	Cranberries	1,000 lb	46.9	48.0	128	216	0.9	0.2	−78	40.1	19.8	−51	16.1	6.4	−60
DE	Broilers	1,000 lb	4.8	4.8	148	260	1.0	0.3	−70	46.4	23.9	−48	18.7	7.6	−59
IL	Corn	1,000 bu	22.2	19.3	1,370	2,150	9.5	2.2	−77	429.5	197.2	−54	172.8	63.2	−63
IL	Hogs	10 head	12.3	11.3	235	845	1.6	0.8	−50	73.7	77.5	+5	29.6	24.8	−63
IL	Soybeans	1,000 bu	18.7	16.5	2,900	6,680	20.1	6.7	−67	909.1	612.8	−33	365.7	196.5	−46
IL	Sulfuric acid	10 tons	5.9	4.6	195	349	1.4	0.4	−71	61.1	32.0	−48	24.7	10.3	−58
MN	Flaxseed	1,000 bu	14.9	20.4	2,450	5,750	17.0	5.8	−66	768.0	527.5	−31	309.0	169.1	−45
MN	Rye	1,000 bu	7.5	10.0	1,150	2,640	8.0	2.7	−66	360.5	242.2	−33	145.0	77.6	−46
MN	Barley	1,000 bu	5.9	13.4	1,140	2,300	7.9	2.3	−71	357.4	211.0	−41	143.8	67.6	−53
GA	Broilers	1,000 lb	14.6	14.2	123	260	0.9	0.3	−67	38.6	23.9	−38	15.5	7.6	−51
VA	Sulfuric acid	10 tons	3.4	2.1	203	401	1.4	0.4	−72	63.6	36.8	−42	25.6	11.8	−54

Source: Calculated from production data in the U.S. Department of Agriculture, *Agricultural Statistics*, 1971, 1972, 1979; and the Bureau of the Census, Industry Division, *Current Industrial Reports* M28A(71)-1 (September 1972), and MA28B(78)-1 (December 1979). Agricultural prices from OBE, U.S. Department of Commerce, *Business Statistics* (1971); *Survey of Current Business* (September 1979); and U.S. Department of Agriculture, *Fats and Oil Situation* (November 1971; February 1979). Energy data from the American Petroleum Institute, *Basic Petroleum Data Book* (January 1980); and Bureau of Mines, *Minerals Yearbook, 1970* (1972). The 1978 coal price was estimated by the Regional Research Institute, West Virginia University. Energy prices were as follows: natural gas, 1970=14.4¢; 1978=99.6¢; crude oil, 1970=$3.19, 1978=$10.90; coal, 1970=$7.93, 1978=$34.00, as cited in William H. Miernyk, "The Differential Effects of Rising Energy Prices on Regional Income and Employment," a paper prepared for the conference.

Table A–3. Residential, Industrial, and Commercial Electric Energy and Natural Gas Prices by State, for Selected Years

| | Electrical energy ($/thousand kWh) | | | | | | Natural gas ($/million Btu) | | | | | |
| | 1970 | | | 1976 | | | 1970 | | | 1978 | | |
State[a]	Res.	Ind.[b]	Comm.[c]	Res.	Ind.[b]	Comm.[c]	Res.	Ind.	Comm.	Res.	Ind.	Comm.
U.S.	21.03	9.48	20.11	34.50	20.68	34.58	1.06	0.38	0.81	2.53	1.91	2.28
ME	26.36	11.23	25.84	34.90	19.42	35.13	2.89	1.42	1.92	5.40	3.33	4.10
NH	26.91	13.35	28.01	45.58	29.61	48.85	1.75	0.98	1.63	3.41	2.99	3.19
VT	21.67	14.43	22.47	40.02	25.86	37.01	1.49	0.48	1.70	3.61	3.07	3.22
MA	27.89	15.57	25.18	49.70	37.35	47.00	1.87	1.00	1.51	4.07	2.69	3.53
RI	27.39	15.44	22.16	48.15	35.99	43.01	1.76	0.71	1.48	4.01	2.75	3.72
CT	23.39	13.65	22.08	44.68	33.15	41.74	1.93	0.99	1.69	4.42	3.06	3.74
NY	28.66	11.20	26.14	53.72	25.52	55.62	1.45	0.77	1.32	3.53	2.47	2.94
NJ	25.41	12.44	23.55	52.71	33.15	48.26	1.62	0.64	1.34	3.80	2.82	3.56
PA	23.21	11.34	20.88	41.20	25.72	37.98	1.17	0.57	0.94	2.81	2.21	2.64
MD	22.78	12.13	21.44	40.55	25.25	41.29	1.43	0.68	1.26	3.54	2.46	3.12
DC	d	d	d	d	d	d	1.44	0.81	0.74	3.46	2.70	3.03
WV	20.80	8.39	18.39	35.85	21.81	34.21	0.87	0.47	0.68	2.32	1.90	2.11
DE	24.44	9.90	20.66	45.26	28.91	41.35	1.50	0.53	1.13	3.94	2.46	3.63
VA	19.82	9.84	17.92	35.08	22.89	34.23	1.42	0.51	1.05	3.31	2.17	2.83
KY	19.00	6.89	18.46	26.24	15.65	16.68	0.84	0.47	0.69	2.02	1.62	1.91
TN	10.83	6.53	14.80	23.00	17.19	27.19	0.89	0.36	0.72	1.98	1.67	2.09
AL	14.99	7.16	16.98	27.58	18.31	30.28	1.10	0.32	0.69	2.73	1.66	2.17
MS	16.41	9.40	17.93	21.32	23.81	29.76	0.88	0.28	0.61	2.37	1.64	2.06
NC	17.68	8.80	15.28	32.94	21.44	29.76	1.29	0.49	1.07	3.03	2.37	2.88
SC	18.30	7.71	16.33	33.06	19.34	29.50	1.34	0.43	0.92	2.93	1.65	2.17
GA	16.83	9.31	18.71	29.33	23.07	35.62	0.98	0.36	0.71	2.71	1.71	1.99
FL	19.79	11.37	21.42	36.65	27.06	37.82	1.78	0.38	1.05	3.33	1.38	2.52
OH	22.69	9.27	20.72	35.64	17.84	33.75	0.89	0.54	0.76	2.43	2.05	2.24
IN	21.29	11.24	20.78	30.20	19.91	29.40	1.05	0.47	0.86	2.18	1.67	2.00
IL	25.88	11.38	22.88	38.58	22.49	37.64	1.02	0.46	0.74	2.43	2.08	2.22
MI	22.70	11.94	23.18	38.88	27.03	40.26	1.00	0.54	0.83	2.30	1.99	2.21
WI	21.90	13.50	23.89	34.37	23.45	33.85	1.21	0.53	0.97	2.58	2.00	2.27
MN	23.70	13.32	25.90	33.29	23.65	35.17	1.08	0.40	0.81	2.50	1.69	2.26

(continued)

Table A–3 *(continued)*.

| | Electrical energy ($/thousand kWh) | | | | | | Natural gas ($/million Btu) | | | | | |
| | 1970 | | | 1976 | | | 1970 | | | 1978 | | |
State[a]	Res.	Ind.[b]	Comm.[c]	Res.	Ind.[b]	Comm.[c]	Res.	Ind.	Comm.	Res.	Ind.	Comm.
NM	26.44	11.00	20.03	35.53	20.84	27.69	0.84	0.28	0.60	2.38	1.61	2.03
TX	20.47	8.01	17.22	31.16	18.48	27.65	0.92	0.23	0.55	2.57	2.13	2.32
OK	24.07	9.98	19.16	29.93	17.63	26.24	0.80	0.27	0.53	1.88	1.48	1.66
AR	22.13	8.89	20.18	33.94	21.52	32.24	0.78	0.27	0.54	1.67	1.33	1.43
LA	21.37	7.95	19.85	26.70	13.29	27.52	0.75	0.23	0.49	2.23	1.65	1.91
NE	20.16	10.92	16.08	29.51	18.60	26.57	0.86	0.32	0.66	1.96	1.33	1.67
KS	23.26	11.18	19.60	31.83	20.52	29.50	0.69	0.26	0.60	1.60	1.38	1.32
IA	25.14	12.36	24.90	36.10	22.24	36.80	0.96	0.36	0.68	2.30	1.65	2.00
MO	25.52	12.80	23.01	35.13	23.55	34.45	0.93	0.36	0.71	2.27	1.43	2.00
MT	21.32	4.24	19.41	22.34	6.09	20.62	0.84	0.34	0.62	2.05	1.69	1.78
WY	24.40	10.32	17.06	24.14	12.02	17.73	0.65	0.24	0.48	2.01	1.27	1.82
CO	25.10	11.19	19.87	32.64	19.24	28.69	0.71	0.26	0.58	1.90	1.34	1.72
ND	25.32	19.00	22.64	30.83	26.69	28.50	1.05	0.44	0.72	2.32	2.14	1.89
SD	25.19	15.07	26.66	31.51	21.58	34.58	1.04	0.33	0.71	2.12	1.47	1.78
UT	21.72	11.98	18.66	31.70	20.03	26.15	0.70	0.28	0.66	1.97	1.32	1.15
CA	21.21	9.27	17.48	35.42	24.33	33.38	0.93	0.37	0.67	1.90	2.17	2.20
NV	14.47	6.85	15.69	28.41	19.54	30.38	1.31	0.41	0.86	2.46	1.64	2.40
AZ	22.70	11.38	18.88	39.02	24.60	35.23	1.14	0.38	0.66	2.97	1.49	2.19
HI	26.68	14.67	32.71	47.66	32.21	54.97	—	—	—	9.03	7.15	7.27
WA	10.13	3.11	11.56	12.68	4.51	14.08	1.29	0.38	1.04	3.29	2.36	2.86
OR	11.86	4.03	12.37	18.57	8.01	17.22	1.44	0.41	1.19	3.67	2.42	3.20
ID	15.61	5.87	13.94	18.34	10.16	17.90	1.34	0.43	0.97	3.37	2.17	2.84
AK	30.14	17.13	31.18	34.49	33.11	32.46	1.42	0.52	0.83	1.81	0.89	1.40

Source: Calculated from data in various issues of *Gas Facts, A Statistical Record of the Gas Utility Industry,* (Arlington, Va., American Gas Association) tables 74, 87 (1971), table 94 (1979); Edison Electric Institute, *Statistical Yearbook of the Electric Utility Industry* (New York: Edison Institute, 1971 and 1977) tables 22S and 36S, as cited in William H. Miernyk, "The Differential Effects of Rising Energy Prices on Regional Income and Employment," a paper prepared for this conference.

[a] States are grouped by Standard Federal Region.

[b] Assumed to be represented by "large light and power" customer classification.

[c] Assumed to be represented by "small light and power" customer classification.

[d] Included with MD.

APPENDIX A

Appendix B
List of conference participants and commissioned papers

**Conference Participants and
Their Affiliations at the Time of the Conference**

Jan Acton, *The Rand Corporation*
Jodie Allen, *U.S. Department of Labor*
Gar Alperovitz, *National Center for Economic Alternatives*
Paul W. Barkley, *Washington State University*
Harold Beebout, *Mathematica Policy Research, Inc.*
Kenneth Bowler, *U.S. House of Representatives Subcommittee on Public Assistance and
    Unemployment Compensation*
Ronald Brunner, *University of Michigan*
Anne P. Carter, *Brandeis University*
Emery N. Castle, *Resources for the Future*
Letitia Chambers, *U.S. Senate Committee on Labor and Human Resources*
Alan L. Cohen, *U.S. Department of Health and Human Services*
Mark N. Cooper, *Consumer Energy Council of America*
Ellis Cose, *Detroit Free Press*
William A. Darity, Jr., *University of Texas, Austin*
Brady J. Deaton, *Virginia Polytechnic Institute and State University*
Otto Doering, *Purdue University*

Joseph M. Dukert, *Consultant*
Edward R. Fried, *The Brookings Institution*
Robert F. Hemphill, Jr., *Tennessee Valley Authority*
Wayne Hoffman, *The Urban Institute*
Kevin Hollenbeck, *Urban Systems, Inc.*
Ronald Johnson, *Federal Reserve Board*
John Korbel, *U.S. Department of Energy, Office of Policy and Evaluation*
Hans H. Landsberg, *Resources for the Future*
Sar A. Levitan, *George Washington University*
Glenn Loury, *University of Michigan*
Bruce MacLaury, *The Brookings Institution*
Eberhard Meller, *International Energy Agency, Organisation for Economic Co-operation and Development*
William H. Miernyk, *West Virginia University*
Milton Morris, *Joint Center for Political Studies*
Gerald Peabody, *U.S. Department of Energy, Energy Information Administration*
Joseph Pechman, *The Brookings Institution*
Frank Potter, *U.S. House of Representatives, Subcommittee on Energy and Power*
Mary Procter, *Office of Technology Assessment*
Larry Ruff, *Brookhaven National Laboratory*
Milton Russell, *Resources for the Future*
Thomas C. Schelling, *Harvard University*
Kenneth A. Small, *Princeton University*
Donald M. Smith, *Staff member, Office of Senator Gary Hart*
Marvin N. Smith, *The Brookings Institution*
Herbert Stein, *University of Virginia and the American Enterprise Institute*
Rodney Stevenson, *University of Wisconsin*
Michael Telson, *U.S. House of Representatives Budget Committee*
Edward Thompson, *The Environmental Action Foundation*
Randall Weiss, *U.S. Congress, Joint Committee on Taxation*
Charles Zielinski, *New York State Public Service Commission*

List of Commissioned Papers

"The Distribution of Household Energy Expenditures and the Impact of High Prices" by Harold Beebout, Gerald Peabody, and Pat Doyle

"Energy Assistance Schemes: Review, Evaluation, and Recommendations" by Alan L. Cohen and Kevin Hollenbeck

"Burden Allocation and Electric Utility Rate Structures: Issues and Options in the TVA Region" by Robert F. Hemphill and Ronald L. Owens

"The Political Economy of U.S. Energy and Equity Policy" by William A. Darity, Jr., Ronald Johnson, and Edward Thompson

"Energy Prices and Real Income Distribution: The Urban Sector" by Kenneth A. Small

"The Impacts of Rising Energy Prices in Rural Areas" by Paul W. Barkley

"The Differential Effects of Rising Energy Prices on Regional Income and Employment" by William H. Miernyk

"The Equity Issue in Europe and Japan" by Eberhard Meller

Panel Session: "Perspectives on Coping with Massive Price Shifts," Remarks by Gar Alperovitz, Thomas C. Schelling, and Herbert Stein

Index

Note: Page numbers followed by *n* indicate that entry appears in a chapter note.

Less-developed countries, 14*n*
Levitan, Sar A. 37*n*
Lifeline utility rates, 53, 54, 61–66, 78*n*
Liquefied petroleum gas (LPG), 21, 23, 24, 93
Lovins, Amory, 9
Lower Living Standard, 19, 66, 69
LPG. *See* Liquefied petroleum gas.

Macroeconomic problems, 9
Managerial class, 3
Manufacturing. *See* Industries.
Marxism and energy issues, 3, 13
Mass transit, 56*n*
Miernyk, William H., 42, 44, 56*n*, 94*n*, 96*n*
Migration. *See* Demographic changes.
Mobility, 40

National Center for Economic Alternatives, 9
National Energy Conservation Act of 1978, 79*n*
National Interim Energy Consumption Survey, 26–27, 28, 29, 31*n*, 32, 47, 48, 49, 68, 78*n*
National transportation survey, 38*n*
Natural gas, 8, 21, 23, 37*n*, 93
  heating, 47
  price controls on, 46
  prices, 23, 45–46, 49, 52, 53, 54, 56*n*, 74, 94–95
Natural Gas Policy Act of 1978, 47, 52–53
Natural resources, nonfuel, 45
NIECS. *See* National Interim Energy Consumption Survey.
Nuclear power, 41*n*, 55, 91*n*

Oak Ridge National Laboratories, 56*n*
Office of Economic Opportunity, 79*n*
  See also Community Services Administration.
Office of Technology Assessment, 54
Offshore resources, 55
Oil embargo, 7, 33
Oil prices, 1, 7, 23, 65

Oil Windfall Profits Tax. *See* Windfall Profits Tax.
Organization of Petroleum Exporting Countries, (OPEC), 10, 65, 67
Owens, Ronald L., 61

Peabody, Gerald, 31*n*, 37*n*, 48, 93*n*
Peak usage patterns, 57
Pipeline delivery, costs of, 23
Pollution and shale oil production, 80*n*
  See also Environmental effects.
Poor. *See* Poverty.
Population. *See* Census Bureau; Demographic changes.
Poverty, 7, 10, 34
  and Consumer Price Index, 12–13
  definition of, 18–19
  energy costs, 2, 3, 34–37, 68
  guaranteed energy supply, 63
  index, of Social Security Administration, 19, 84
  inner city areas, 51–52
  level, 34–35
  measurement, 86
  poverty line, 31*n*
  standards of, 66, 69–70, 76, 79*n*, 84
  utility rates, 61–66
  See also Assistance programs.
President's Commission for a National Agenda for the Eighties, 40, 55*n*
Price controls, 45–46, 60, 61
Price Index, for the poor, 12–13, 89
Procter, Mary, 54
Property values and energy premiums, 50–51
Psychological factors
  income, 11
  migration, 44
  rural life, 49
Public opinion, and energy policy, 88–89
Rand Corporation, 35, 38*n*, 78*n*
Reagan administration
  block cash grants, 63
  decontrol of oil prices, 7

Reagan administration *(continued)*
  energy assistance, 76–77
  free enterprise zones, 56
Regional factors, 23, 54, 89
  assistance programs, 73
  business shifts, 44
  employment patterns, 44
  guaranteed supply, 63
  income trends, 47
  natural gas rates, 54
  residential expenditures, 48
  weather, 48–49, 83
Regional Research Institute, 94*n*
Rent control, 67
Residential Energy Consumption Survey (RECS), 26
Residential energy use
  direct energy expenditure, 35, 82
  energy consumption, 8, 9, 24–29, 93
  energy expenditures, 27, 38*n*, 93
  energy purchases, 26
  energy use in, 25
  expenditures, 69, 84, 93
  heating 28, 82
  income, 21, 22*n*
  rural, 56
  socioeconomic characteristics, 30, 31
  transportation costs, 32
  weatherization, 71. *See also* Weatherization.
Revenue sharing, 3, 75, 76, 79*n*
Reynolds, Reid T., 56*n*
Ruff, Larry, 14
Rural areas, 49, 68

Schelling, Thomas C., 2, 89
Security premium on imported oil, 7
Self-sufficiency, 10, 40–43
Severance taxes, 45, 75
SFR (Standard Federal Regions), 44
Shale oil, 80*n*
Sheltered enterprises, 44
Slurry pipelines, 55
Smith, Donald M., 46, 56*n*